Essential Statistics

Essential Statistics

Third edition

D.G. Rees

Senior Lecturer in Statistics
Oxford Brookes University
UK

CHAPMAN & HALL

London · Glasgow · Weinheim · New York · Tokyo · Melbourne · Madras

Published by Chapman & Hall, 2–6 Boundary Row, London SE1 8HN, UK

Chapman & Hall, 2–6 Boundary Row, London SE1 8HN, UK

Blackie Academic & Professional, Wester Cleddens Road, Bishopbriggs, Glasgow G64, 2NZ, UK

Chapman & Hall GmbH, Pappelallee 3, 69469 Weinheim, Germany

Chapman & Hall USA, One Penn Plaza, 41st Floor, New York NY 10119, USA

Chapman & Hall Japan, ITP-Japan, Kyowa Building, 3F, 2-2-1 Hirakawacho, Chiyoda-ku, Tokyo 102, Japan

Chapman & Hall Australia, Thomas Nelson Australia, 102 Dodds Street, South Melbourne, Victoria 3205, Australia

Chapman & Hall India, R. Seshadri, 32 Second Main Road, CIT East, Madras 600 035, India

First edition 1985
Second edition 1989
Reprinted 1990, 1991 (twice), 1992, 1994
Third edition 1995

Typeset in 10/12 Times by Best-set Typesetter Ltd., Hong Kong
Printed in Great Britain by T.J. Press Ltd, Cornwall

ISBN 0 412 61280 1

A catalogue record for this book is available from the British Library

∞ Printed on permanent acid-free text paper, manufactured in accordance with ANSI/NISO Z39.48-1992 and ANSI/NISO Z39.48-1984 (Permanence of Paper).

Contents

Preface

For this edition, the second edition has been completely reviewed and appropriately revised and rewritten. In addition, there are a number of new sections. For example, Minitab applications have been included within each chapter as they arise, rather than in a special chapter at the end of the book. A short introduction to Minitab is given in an Appendix. A new data set (40 cases, 6 variables) has been introduced as a basis for many of the examples in the text. There are new sections on Venn diagrams, the F test for the equality of two variances, the Fisher exact text, the χ^2 trend test and the Shapiro–Wilk test for normality. Some methods applicable to grouped data, for example the mean and standard deviation, have been omitted in this new edition, since it can be assumed that, nowadays, all data are initially input case by case to a computer or calculator. The worksheets at the end of each chapter have also been reviewed and revised. Detailed solutions are again provided, and there is a completely new multiple-choice test.

There is a view that the advent of the statistical computer package has dispensed with the need for the calculator, statistical formulae and statistical tables. I do not share this view. I believe that, for a proper and deep understanding of the concepts of statistics and the analysis of statistical data, it is essential to know what the computer or calculator is doing with the data, what assumptions are being made in carrying out an analysis and whether these assumptions are reasonable assumptions, and also the limitations of each method. The computer may take some of the drudgery out of the calculations, but it is not a substitute for careful thought. The reader will find that virtually all the methods described in this book may be performed by hand, i.e. with a calculator using given formulae and tables, and also by computer, i.e. using Minitab. The underlying assumptions and limitations are given and fully discussed.

Finally, I hope that the 'friends of *Essential Statistics*', who have found earlier editions of value, will also like this new edition.

D.G. Rees

Preface to the second edition

The main feature of this new edition is a substantial addition on applications of the interactive statistical computer package, Minitab. This package has become widely used in colleges as an aid to teaching statistics. The new chapter contains over 20 sample programs illustrating how Minitab can be used to draw graphs, calculate statistics, carry out tests and perform simulations. The chapter could act as a primer for first-time Minitab users.

There are also new sections in Chapters 3 and 4 on some aspects of exploratory data analysis. Some changes have been made to the statistical tables. For example, Tables D.1 and D.2 now give cumulative probabilities in terms of '*r* or fewer . . .' instead of '*r* or more . . .'. The tables are now consistent with those adopted by most GCSE examination boards and also with the output from the Minitab **CDF** command for both the binomial and Poisson distributions. For similar reasons Table D.3(a) now gives the cumulative distribution function for normal distribution, i.e. areas to the left of various values of z. Another change is that the conditions for the use of the normal approximation to the binomial have been brought into line with accepted practice. There are other minor changes too numerous to list here.

I am grateful for the opportunity to update and enhance the successful first edition. Many thanks to all those who have expressed their appreciation of *Essential Statistics* as a course text or who have made helpful suggestions for improvements.

Preface to the first edition

TO THE STUDENT

Are you a student who requires a basic statistics text-book? Are you studying statistics as part of a study of another subject, for example one of the natural, applied or social sciences, or a vocational subject? Do you have an O-level or GCSE in mathematics or an equivalent qualification? If you can answer 'yes' to all three questions I have written this book primarily for you.

The main aim of this book is to encourage and develop your interest in statistics, which I have found to be a fascinating subject for over twenty years. Other aims are to help you to:

1. Understand the essential ideas and concepts of statistics.
2. Perform some of the most useful statistical methods.
3. Be able to judge which method is the most appropriate in a given situation.
4. Be aware of the assumptions and pitfalls of the methods.

Because of the wide variety of subject areas which require knowledge of introductory statistics, the worked examples of the various methods given in the main part of the text are not aimed at any one subject. In fact they deliberately relate to methods which can be applied to 'people data' so that every student can follow them without specialist knowledge. The end-of-chapter Worksheets, on the other hand, relate to a wide variety of subjects to enable different students to see the relevance of the various methods to their areas of special interest.

You should tackle each worksheet before proceeding to the next chapter. To help with the necessary calculations you should be, or quickly become, familiar with an electronic hand calculator with the facilities given below.[†] (These facilities are now available on most scientific cal-

[†] *Calculators* The minimum requirements are: a memory, eight figures on the display, a good range of function keys (including square, square root, logarithm, exponential, powers, factorials) and internal programs for mean and standard deviation.

culators.) Answers and partial solutions are given to all the questions on the worksheets. When you have completed the whole book (except for the sections marked with an asterisk (*), which may be omitted at the first reading), a multiple-choice test is also provided, as a quick method of self-assessment.

TO THE TEACHER OR LECTURER

This book is not intended to do away with face-to-face teaching of statistics. Although my experience is that statistics is best taught in a one-to-one situation with teacher and student, this is clearly not practical in schools, colleges and polytechnics where introductory courses in statistics to non-specialist students often demand classes and lectures to large groups of students. Inevitably these lectures tend to be impersonal.

Because I have concentrated on the essential concepts and methods, the teacher who uses this book as a course text is free to emphasize what he or she considers to be the most important aspects of each topic, and also to add breadth or depth to meet the requirements of the particular course being taught.

Another advantage for the teacher is that, since partial solutions are provided to all the questions on the worksheets, students can attempt these questions with relatively little supervision.

WHAT THIS BOOK IS ABOUT

After introducing 'Statistics as a science' in Chapter 1 and statistical notation in Chapter 2, Chapters 3 and 4 deal with descriptive or summary statistics, while Chapters 5, 6 and 7 concentrate on probability and four of the most useful probability distributions.

The rest of the book comes broadly under the heading of statistical inference. After discussing sampling in Chapter 8, two branches of inference – confidence interval estimation and hypothesis testing – are introduced in Chapters 9 and 10 by reference to several 'parametric' cases. Three non-parametric hypothesis tests are discussed in Chapter 11.

In Chapters 12 and 13 association and correlation for bivariate data are covered. Simple linear regression is dealt with Chapter 14 and χ^2 goodness-of-fit tests in Chapter 15.

I have attempted throughout to cover the concepts, assumptions and pitfalls of the methods, and to present them clearly and logically with the minimum of mathematical theory.

Acknowledgements

The quotations given at the beginning of Chapters 1, 2, 3, 4, 8, 10, 11, and 13 are taken from a very interesting book on diseases and mortality in London in the eighteenth century. I would like to thank Gregg International, Amersham, England for permission to use these quotations from *An Arithmetical and Medical Analysis of the Diseases and Mortality of the Human Species* by W. Black (1973).

Acknowledgements for the use of various statistical tables are given in Appendix D.

Thanks also to all the colleagues and students who have influenced me, and have therefore contributed indirectly to this book. Most of all I am grateful to my wife, Merilyn, for her support and encouragement throughout.

What is statistics?

> Authors . . . have obscured their works in a cloud of figures and calculation: the reader must have no small portion of phlegm and resolution to follow them throughout with attention: they often tax the memory and patience with a numerical superfluity, even to a nuisance.

1.1 STATISTICS AS A SCIENCE

You may feel that the title of this chapter should be 'What are statistics?', indicating the usual meaning of statistics as numerical facts or numbers. So, for example, the unemployment statistics are published monthly giving the number of people who have received unemployment benefit during the month. However, in the title of this chapter the singular noun 'statistics' is used to mean the science of collecting and analysing data, where the plural noun 'data' means numerical or non-numerical facts or information.

We may collect data about 'individuals', that is individual people or objects. There may be many characteristics which vary from one individual to another. We call these characteristics **variables**. For example, individual people vary in height and employment status, and so height and employment status are variables.

Let us consider an example of some data which we might wish to analyse. Suppose our variable of interest is the height of students at the start of their first year in higher education. We would expect these heights to vary. We could start by choosing one college from all the colleges of higher education, we could choose 40 first-year students from the college's enrolment list, and we could measure the heights of these students – see column 3 of Appendix A – and calculate the average height. There are many other ways of collecting and analysing such data. Indeed, this book is about how surveys like this should be conducted, and clearly they cannot be discussed now in detail. But it is instructive to ask some of the questions which need to be considered before such a survey is carried out.

The most important question is: 'What is the purpose of the survey?' The answer to this question will help us to answer other questions. Is it better to choose all 40 students from one college or a number from each of a number of colleges? How many students should be selected altogether, and how many from each of the chosen colleges? How do we select a given number of students from a college's enrolment list? What do we do if a selected student refuses to co-operate in the survey? How do we allow for known or suspected differences between male and female student heights? How accurately should the heights be measured? Does the average height of the students selected for the survey tell us all we need to know about their heights? How can we relate the average height of the selected students to the average height of all students at the start of their first year in higher education?

1.2 TYPES OF STATISTICAL DATA

Before we look at how data may be collected and analysed, we will consider the different types of statistical data we may need to study. As stated in the Preface to the First Edition, the main part of this book will be concerned with 'people data', so the following list gives some of the variables which may be collected from people, e.g. the 40 students listed in Appendix A:

Sex
Height
Number of brothers and sisters (siblings)
Distance from home to Oxford
Type of degree
A-level count

Some of these variables are **categorical**, that is the 'value' taken by the variable is a non-numerical category or class. An example of a categorical variable is sex, with categories male and female. Some variables are quantifiable, that is they can take numerical values. These numerical variables can be further classified as being either continuous, discrete or ranked using the following definitions:

A **continuous** variable can take any value in a given range.
A **discrete** variable can take only certain distinct values in a given range.
A **ranked** variable is a categorical variable in which the categories imply some order or relative position.

Table 1.1 Examples of types of statistical data

Name of variable	*Type of variable*	*Likely range of values or list of categories*
Sex	Categorical	Male, female
Height	Continuous	100 to 200 cm
Number of siblings	Discrete	0, 1, . . ., 10
Distance home to Oxford	Continuous	0 to 500 km
Type of degree	Categorical	BA, BSc
A-level count	Discrete	0, 1, 2, . . . , 50

Example

Height is an example of a continuous variable since an individual adult human being may have a height anywhere in the range 100 to 200 cm, say. We can usually decide that a variable is continuous if it is measured in some units.

Example

Number of brothers and sisters (siblings) is an example of a discrete variable, since an individual human being may have 0, 1, 2, . . . siblings, but cannot have 1.43, for example. We can usually decide that a variable is discrete if it can be counted.

Example

Birth order is an example of a ranked variable, since an individual human being may be the first-born, second-born, etc., into a family, with corresponding birth order of 1, 2, etc.

Table 1.1 shows the results of applying similar ideas to all the variables in the above list.

You may feel that the distinction between the continuous and the discrete variable is, in practice, not as clear-cut as stated above. For example, most people give their age as a whole number of years, so that age appears to be a discrete variable which increases by one at each birthday. The practice of giving one's age approximately, for whatever reason, does not alter the fact that age is fundamentally a continuous variable.

Now try Worksheet 1.

WORKSHEET 1: TYPES OF STATISTICAL DATA

For the following decide whether the variable is continuous, discrete, ranked or categorical. Give a range of likely values or a list of categories.

The value or category of the variable varies from one 'individual' to another. The individual may or may not be human, as in the first question where the individual is 'lightbulb'. Name the individual in each case.

1. The number of hours of operation of a number of 100 W lightbulbs.
2. The number of current-account balances checked by a firm of auditors each year.
3. The present cost of bed and breakfast in three-star London hotels.
4. The number of rooms with bathroom in three-star London hotels.
5. The type of occupation of adult males.
6. The number of failures per 100 hours of operation of a large computer system.
7. The number of hours lost per 100 hours due to failures of a large computer system.
8. The number of cars made by a car company each month.
9. The position of the British entry in the annual Eurovision song contest each year.
10. The annual rainfall in English counties in 1993.
11. The number of earthquakes per year in a European country in the period 1900–1993.
12. The outputs of North Sea oil rigs in 1993.
13. The α-particle count from a radioactive source in 10-second periods.
14. The number of times rats turn right in ten encounters with a T-junction in a maze.
15. The grades obtained by candidates taking A-level mathematics.
16. The colour of people's hair.
17. The presence or absence of a plant species in each square metre of a meadow.
18. The reaction time of rats to a stimulus.
19. The number of errors per page of a balance sheet.
20. The yield of tomatoes per plant in a greenhouse.
21. The constituents found in core samples when drilling for oil.
22. The percentage hydrogen content of gases collected from samples near to a volcanic eruption.
23. The political party people vote for in an election.

Solutions to this and other worksheets are given in Appendix C.

Some statistical notation

I have corrected several errors of
preceding calculators . . .

It is not necessary for you to master all the notation in this chapter before you proceed to Chapter 3. However, references to this notation will be made in later chapters within the context of particular statistical methods. Worksheet 2 which follows this chapter is intended to help you to use your calculator and become familiar with the notation.

2.1 Σ

The symbol Σ (the upper-case version of the Greek letter sigma) implies the operation of summing. If x stands for a variable, then Σx means 'sum all the observations of x'. If there are n observations in a sample taken from a population, then we can write:

$$\text{sample mean of } x = \frac{\text{sum of the observed values of } x}{\text{number of observed values}}$$

The formula for this is:

$$\bar{x} = \frac{\Sigma x}{n}$$

We pronounce $\bar{x}$ as 'x bar'. You will find $\bar{x}$ on any scientific calculator, while Minitab simply uses the word 'mean'.

Example

The sample of five coins in my pocket have the following values (p):

1, 2, 2, 5, 100

So we can write:

$$\text{sample mean} = \frac{1 + 2 + 2 + 5 + 100}{5} = 22$$

Using the formula, $n = 5$, $\Sigma x = 110$, $\bar{x} = 110/5 = 22\text{p}$.

Other uses of the Σ notation are Σx^2, $(\Sigma x)^2$, $\Sigma(x - \bar{x})$, defined as follows:

Σx^2 means square the n observations of x and then sum.

$(\Sigma x)^2$ means sum the n observations of x and then square this sum.

$\Sigma(x - \bar{x})$ means subtract the sample mean from each observation of x, and then sum.

Example

Carry out the above operations on the data from the previous example:

$$\Sigma x^2 = 1^2 + 2^2 + 2^2 + 5^2 + 100^2 = 10\,034 \quad \text{units are 'square pence' } (\text{p}^2)!$$

$$(\Sigma x)^2 = (1 + 2 + 2 + 5 + 100)^2 = 110^2 = 12\,100\,\text{p}^2$$

$$\begin{aligned}\Sigma(x - \bar{x}) &= (1 - 22) + (2 - 22) + (2 - 22) + (5 - 22) + (100 - 22)\\ &= -21 - 20 - 20 - 17 + 78\\ &= 0.\end{aligned}$$

Note that $\Sigma(x - \bar{x})$ will always be zero for any set of data.

2.2 FACTORIALS

If n is a positive whole number, then $1 \times 2 \times 3 \times \cdots \times n$ is called **factorial** n and is written $n!$. So

$$n! = 1 \times 2 \times 3 \times \cdots \times n$$

Examples

$$\begin{aligned}3! &= 1 \times 2 \times 3 = 6\\ 5! &= 1 \times 2 \times 3 \times 4 \times 5 = 120\\ 1! &= 1\end{aligned}$$

Try these on your calculator.

In addition to the above definition of factorial n for positive whole numbers, factorial 0 is defined as 1, so $0! = 1$. Try this on your calculator as well.

Remember that factorials for any other numbers are not defined. So $-5!$ and $2.3!$ are not defined and hence are meaningless.

Factorials will be used initially in this book in calculating binomial probabilities in Chapter 6.

2.3 x^y

To find the 'power y of any number x' you need the x^y button on your calculator.

Example

$(0.6)^4$ implies $x = 0.6$, $y = 4$. $(0.6)^4 = 0.6 \times 0.6 \times 0.6 \times 0.6 = 0.1296$. Check this on your calculator using the x^y button.
$(0.6)^0$ implies $x = 0.6$, $y = 0$. $(0.6)^0 = 1$. Check this on your calculator.

The x^y button is useful in calculating binomial probabilities (Chapter 6).

2.4 e^x

The letter e in mathematics and on your calculator stands for the number 2.718 (approx.). We need to be able to obtain values of e^x in Chapter 6 in the calculation of Poisson probabilities.

Example

$$e^1 = e = 2.718$$
$$e^{-2} = 0.1353$$
$$e^0 = 1$$

Try these on your calculator.

2.5 DECIMAL PLACES AND SIGNIFICANT FIGURES

Calculators produce many figures on the display and it is tempting to write them all down. You will learn by experience how many figures are meaningful in an answer. For the moment, concentrate on giving answers to a stated number of decimal places or significant figures.

Use the idea that, for example, 3 decimal places (dps) means write three figures only to the right of the decimal point, rounding the third figure (after the decimal point) up if the fourth figure is 5 or more.

Examples

1.6666 to 3 dps is 1.667.
1.6665 to 3 dps is 1.667.
1.6663 to 3 dps is 1.666.

1.67 to 3 dps is 1.670.
167 to 3 dps is 167.000.

The number of significant figures means the number of figures (as you scan from left to right) starting with the first non-zero figure. Round the last significant figure up if the figure immediately to its right is 5 or more. Non-significant figures to the left of the decimal point are written as zeros, and those to the right of the decimal point are omitted.

Examples

26 243 to 3 sig. figs is 26 200.
2624 to 3 sig. figs is 2620.
2626 to 3 sig. figs is 2630.
26.24 to 3 sig. figs is 26.2.
0.2624 to 3 sig. figs is 0.262.
0.002 626 to 3 sig. figs is 0.002 63.

WORKSHEET 2: SOME STATISTICAL NOTATION

1. Check that you are able to work out each of these on your calculator.
   (a) $1.3 + 2.6 - 5.7$
   (b) $10.0 - 3.4 - 2.6 - 1.0$
   (c) $(2.3)(14.6)$
   (d) $(0.009)(0.0274)(1.36)$
   (e) $2.3/14.6$
   (f) $1/0.002\,93$
   (g) $(2.3 + 4.6 + 9.2 + 17.3)/4$
   (h) $28^{0.5}$
   (i) $(0.5)^3$
   (j) $(0.2)^2(0.8)^4$
   (k) $(0.5)^0$
   (l) $(0.2)^{-3}$
   (m) $e^{1.6}$
   (n) $e^{-1.6}$
   (o) $13/\sqrt{(10 \times 24)}$
   (p) $6 - (-0.5)(4)$
   (q) $4!, 1!, 6!, 0!, (-3)!, (2.4)!$

2. Express the answer to Question:
   (a) 1(c) to 1 dp.
   (b) 1(d) to 2 sig. figs.
   (c) 1(e) to 2 sig. figs.
   (d) 1(f) to 4 sig. figs.
   (e) 1(f) to 1 sig. fig.

3. Use the memory facility on your calculator to work out the following:
   (a) $1 + 2 + 3 + 4 + 5 + 6 + 7 + 8 + 9 + 10$
   (b) $(1 + 2 + 3 + 4 + 5)/5$
   (c) $1^2 + 2^2 + 3^2 + 4^2 + 5^2$
   (d) $(1 \times 2) + (3 \times 4) + (5 \times 6)$.

4. For the eight observations of x: 2, 3, 5, 1, 4, 3, 2, 4, find Σx, $\bar{x}$, $(\Sigma x)^2$, Σx^2, $\Sigma(x - \bar{x})$, $\Sigma(x - \bar{x})^2$ and $\Sigma x^2 - \dfrac{(\Sigma x)^2}{n}$.

5. Repeat Question 4 for the five observations of x: 2.3, 4.6, 1.3, 7.2, 2.3.

Summarizing data by tables and by graphical methods

The important data . . . are condensed, classed, and arranged into concise tables.

If we collect data it is often a good idea to use tabular and graphical methods to 'explore' the data, before we do any calculations. Several examples will be given using all the types of data discussed in Chapter 2. Initially we will concentrate on one-variable data, but later bivariate (two-variable) data will be considered.

3.1 TABLES AND GRAPHS FOR ONE CONTINUOUS VARIABLE

Column 3 of Appendix A gives the heights of a sample of 40 students. These heights may be rewritten as shown in Table 3.1.

We may represent these data graphically in several different ways, for example,

(a) a **dotplot** (Fig. 3.1),
(b) a **stem and leaf** display (Fig. 3.2),
(c) a **box and whisker plot** (Fig. 3.3).

The interpretation of Fig. 3.1 is relatively straightforward. Each observation is represented by one dot on the scale of the variable, which is height in this case. Looking at the dotplot, we see that the dots are fairly evenly spread across the range 155 to 185, with perhaps a tendency

Table 3.1 List of the heights (cm) of 40 students

183	163	152	157	157	165	173	180	164	160
166	157	168	167	156	155	178	169	171	175
169	168	165	166	164	163	161	157	181	163
157	169	177	174	183	181	182	171	184	179

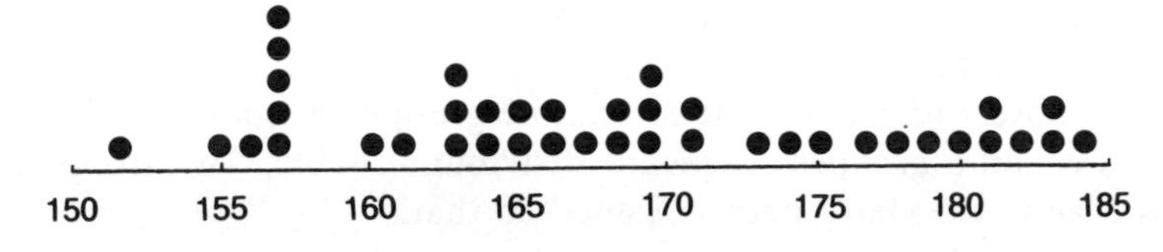

Fig. 3.1 Dotplot for the data in Table 3.1.

```
15 | 2
15 | 5 6 7 7 7 7 7
16 | 0 1 3 3 3 4 4
16 | 5 5 6 6 7 8 8 9 9 9
17 | 1 1 3 4
17 | 5 7 8 9
18 | 0 1 1 2 3 3 4
```

Fig. 3.2 Stem and leaf display for the data in Table 3.1.

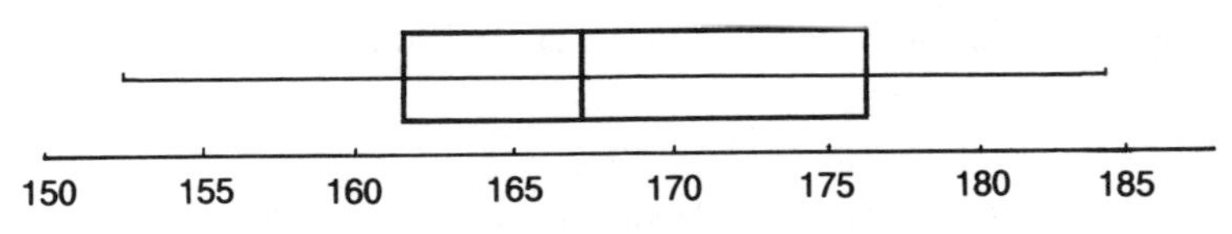

Fig. 3.3 Box and whisker plot for the data in Table 3.1.

to bunch more in the range 163 to 170. The dots are more or less symmetrically distributed about a middle value of about 167 cm.

The stem and leaf display is a way of representing the data in a mixture of a graph and a table. The column of numbers to the left of the vertical line is the stem in Fig. 3.2, while the values to the right of the line are the leaves. The first row in Fig. 3.2 is for observations 150 to 154, while observations from 155 to 159 go in the second row, and so on. Note that the leaves are written down in rank order. You need to turn Fig. 3.2 through 90 degrees to compare its shape with Fig. 3.1. Thc interpretation is similar to that above for the dotplot.

The box and whisker plot is the hardest at this stage to interpret because we do not know what the box (i.e. the rectangle) represents, nor do we know what the whiskers (i.e. the straight line through the box) represent. In fact the mark on the whisker represents the median value for the variable (167.5 for our data), the ends of the wiskers are the minimum (smallest) and maximum (largest) values (152 and 184 respectively for our data), while the points where the whiskers intersect with the box correspond to the lower and upper quartiles (161.5 and 176.5 respectively for our data). We will meet the terms median, lower quartile and upper quartile in Chapter 4.

The 40 observations in Table 3.1 can be grouped as shown in Table 3.2, which is called a **grouped frequency distribution** table. The groups 149.5 to 154.5 and so on have been designed so that:

(a) there are between 5 and 10 groups for smallish data sets (and up to 15 groups for large data sets, e.g. where the total is above 500). If there are too few groups the distribution of the data, i.e. the way they vary, cannot be seen in sufficient detail. If there are too many groups, the table becomes less of a summary.
(b) each observation can fit into one and only one group. In other words, the groups should not overlap. For example, it is clear that 160 goes into the third group, while 159 goes into the second group.
(c) the groups are equally wide, unless there is a good reason for groups of unequal width. In Table 3.2, each group is 5 cm wide (154.5 − 149.5). It is also easier to represent data graphically if the groups are all of the same width. Table 3.2 can be represented graphically in the form of a **histogram** (Fig. 3.4), noting that the mid-points of the groups are 152 (= (149.5 + 154.5)/2), 157 and so on.

Table 3.2 Grouped frequency distribution for the heights (cm) of 40 students

Height	*Number of students (frequency)*
149.5 to 154.5	1
154.5 to 159.5	7
159.5 to 164.5	7
164.5 to 169.5	10
169.5 to 174.5	4
174.5 to 179.5	4
179.5 to 184.5	7
Total	40

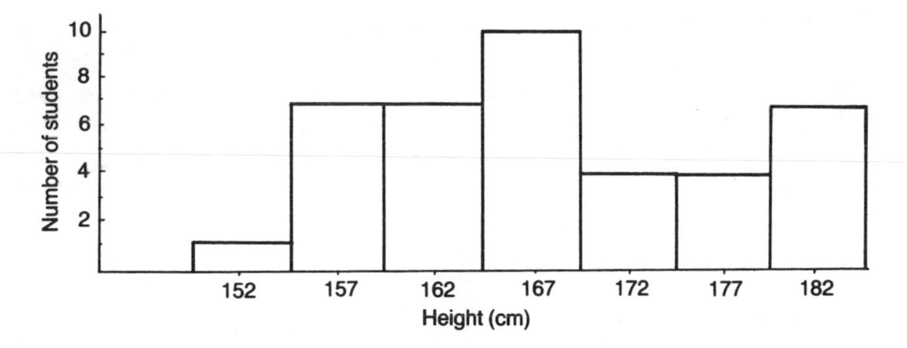

Fig. 3.4 Histogram for the data in Table 3.2.

Note that the vertical axis of the histogram is frequency only if the groups are of equal width, as they are in this example.

The interpretation of Fig. 3.4 has to be the same as that for Fig. 3.2 since they have identical shapes! Try turning Fig. 3.2 through 90 degrees anti-clockwise, and you will see this for yourself.

The next table in this section (Table 3.3) is the **cumulative frequency distribution** table, which we derive from Table 3.2. The values in the height column are group end-points. The table provides information such as '8 students have a height of less than 159.5 cm'.

Table 3.3 can be represented graphically in the form of a **cumulative frequency polygon**. Notice that each row of Table 3.3 gives rise to a point on Fig. 3.5, starting with a cumulative frequency of zero and ending with a cumulative frequency which is the total frequency (40 in the example). It is a common mistake to use group mid-points rather than group end-points in the cumulative frequency polygon.

Table 3.3 Cumulative frequency distribution table for the heights (cm) of 40 students

Height	*Cumulative number of students (cumulative frequency)*
149.5	0
154.5	1
159.5	8 (= 1 + 7)
164.5	15
169.5	25
174.5	29
179.5	33
184.5	40

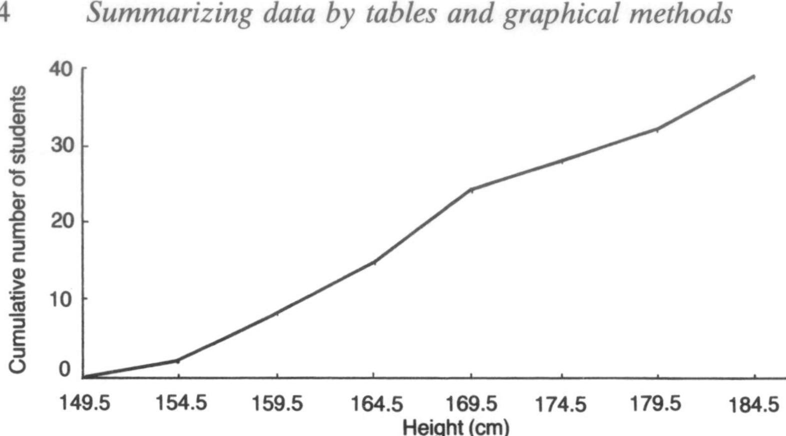

Fig. 3.5 Cumulative frequency polygon for the data in Table 3.3.

Finally, Table 3.4 shows the Minitab commands which will produce a dotplot (like Fig. 3.1), a stem and leaf display (like Fig. 3.2), a box and whisker plot (like Fig. 3.3) and a histogram (like Fig. 3.4). If you are not familiar with Minitab data entry and editing, please refer to Appendix F.

3.2 TABLES AND GRAPHS FOR ONE DISCRETE VARIABLE

The numbers of siblings for 40 students, given in column 4 of Appendix A, are reproduced in Table 3.5. One could draw a dotplot for these data, as we have done in Fig. 3.6. However, a stem and leaf display would not be useful since each observation is only one integer, which cannot be both a stem and a leaf. A box and whisker plot might be useful, but we will wait until Chapter 4 to discuss the terms median, lower and upper quartile for a discrete variable.

In Fig. 3.6 the distribution is clearly bunched around the values 1 and 2, but it is also positively skew ('tail' to the right), because the values to the right of the bunch, i.e. 4, 5 and 9 are not balanced by values to the left of the bunch.

A neat table for the data in Table 3.5 is a grouped frequency distribution table, similar to Table 3.2 (see Table 3.6). This table contains exactly the same information as Fig. 3.6, as does Fig 3.7 which is a line chart. The line chart is the equivalent, for discrete data, of the histogram for continuous data (see Fig. 3.4).

Neither the cumulative frequency distribution table nor the cumulative polygon are useful for discrete data because:

(a) statements like 18 out of 40 students have fewer than 2 siblings do not seem that informative.

Table 3.4 Minitab commands to produce four graphs from the data in Table 3.1

```
MTB> OUTFILE 'HEIGHTS'
MTB> SET C1
DATA> 183 163 152 157 157 165 173 180 164 160
DATA> 166 157 168 167 156 155 178 169 171 175
DATA> 169 168 165 166 164 163 161 157 181 163
DATA> 157 169 177 174 183 181 182 171 184 179
DATA> END
MTB> NAME C1 'HEIGHT'
MTB> PRINT C1
MTB> DOTPLOT C1
MTB> STEM-AND-LEAF C1
MTB> BOXPLOT C1
MTB> HISTOGRAM C1;
SUBC> INCREMENT 5;
SUBC> START 152.
MTB> NOOUTFILE
MTB> STOP
```

Notes

1. The commands OUTFILE and NOOUTFILE create an output file called HEIGHTS.LIS which can be printed. For example, at Oxford Brookes, using a microcomputer network, having left Minitab via the STOP command, you can choose DOS.PROMPT from a menu, and then the command COPY HEIGHTS.LIS PRN results in output to a local printer.
2. The HISTOGRAM command must end with a semicolon if you want to choose the groups yourself, as in Table 3.4. The subcommands INCREMENT 5 and START 152 means that the mid-points of the groups will be 152, 157 (= 152 + 5), and so on.

Table 3.5 The number of siblings for 40 students

1	2	2	3	1	3	1	2	2	3
0	1	0	2	1	1	1	3	5	3
2	4	1	1	3	1	1	2	2	1
2	2	2	1	1	2	1	9	2	1

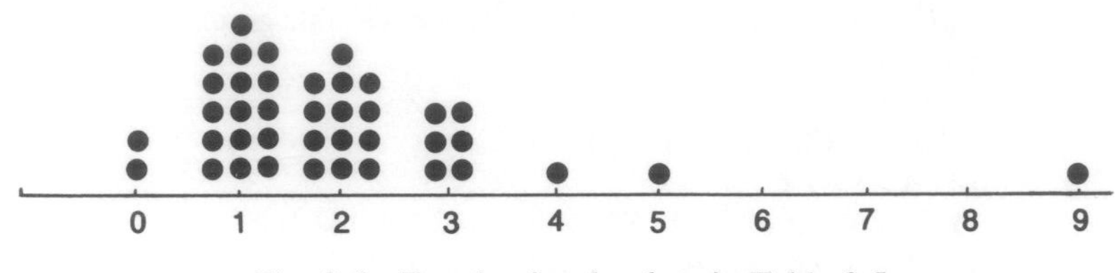

Fig. 3.6 Dotplot for the data in Table 3.5.

Table 3.6 Grouped frequency distribution for the number of siblings of 40 students

Number of siblings	*Number of students (frequency)*
0	2
1	16
2	13
3	6
4	1
5	1
6	0
7	0
8	0
9	1

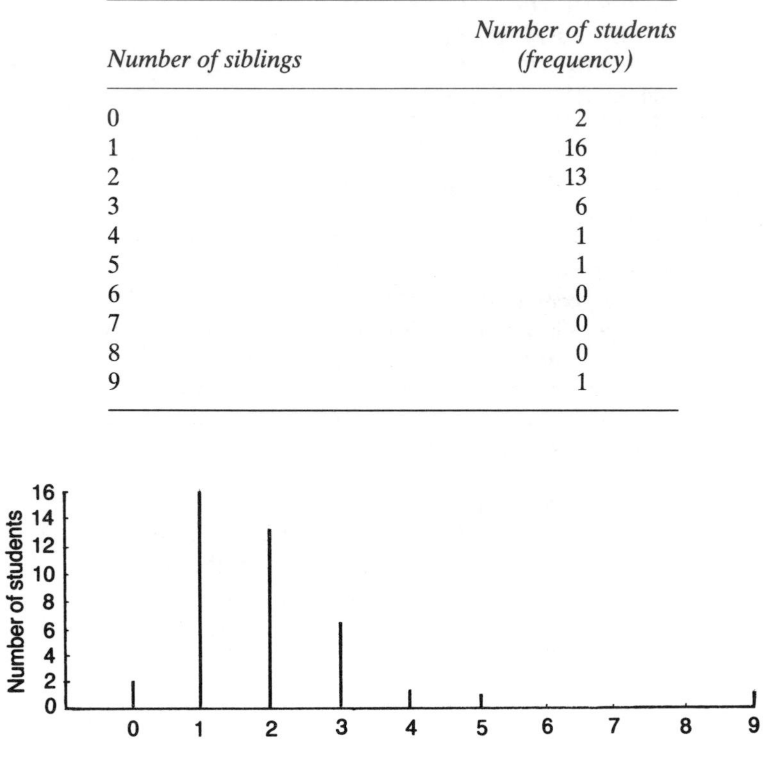

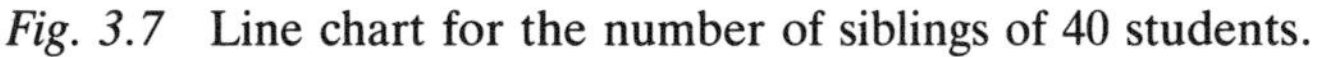

Fig. 3.7 Line chart for the number of siblings of 40 students.

(b) the cumulative frequency polygon would consist of a series of steps rather than a simple set of points joined together.

Minitab does not distinguish between continuous and discrete data, so it is possible to draw (Minitab) graphs for discrete data which may be neither sensible nor useful.

3.3 TABLES AND GRAPHS FOR ONE CATEGORICAL VARIABLE

Column 2 of Appendix A indicates the sex of 40 students, which can be summarized as in Table 3.7. We can draw a **bar chart** of these data, as shown in Fig. 3.8.

Minitab requires all data to be in numerical form, which is why 'male' and 'female' are represented by numbers 1 and 2 respectively in Appendix A. Try the Minitab commands shown in Tablc 3.8.

Table 3.7 A grouped frequency table for the sex of 40 students

Sex	*Number of students (frequency)*
Male	13
Female	27

Fig. 3.8 Bar chart for the data in Table 3.7.

Table 3.8 Minitab commands to produce a histogram of the data in Table 3.7

```
MTB> SET C1
DATA>  1  2  2  2  2  2  1  1  2  2
DATA>  2  2  2  2  2  2  1  2  2  1
DATA>  1  2  2  2  2  2  2  2  1  2
DATA>  2  2  2  2  1  1  1  1  1  1
DATA> END
MTB> NAME C1 'SEX'
MTB> HISTOGRAM C1
```

3.4 TABLES AND GRAPHS FOR TWO-VARIABLE DATA

These will not be discussed in detail here, but a few specific examples will be given because they will be referred to in later chapters.

When both variables are categorical, the frequencies of the various cross-categories may be displayed in a two-way or contingency table (see Table 3.9) and graphically in a pictogram (see Fig. 3.9). An analysis of this type of data is discussed in Chapter 12, for example to answer the question 'Is there some association between sex and type of degree, or are they independent?'.

When both variables are continuous, the data may be displayed in two columns or rows (see columns 3 and 5 of Appendix A for the variables 'height' and 'distance'). A useful graphical method in this case is the scatter diagram (see Fig. 3.10). An analysis of this type of data is discussed in Chapters 13 and 14, for example to answer questions like 'Is there any correlation between the height of a student and the distance he/she lives from Oxford?', and 'Can we predict one of these variables from the other with reasonable accuracy?'. A Minitab program to obtain this scatter diagram would be as shown in Table 3.10.

When one variable is continuous and the other is categorical (with only a few categories) or discrete (with only a few possible values), a good idea is to draw two (or more) dotplots. For example, to compare the heights of male and female students using the data in columns 2 and 3 of Appendix A, we obtain Fig. 3.11.

Table 3.11 gives a Minitab program for the same plots. Note the use of the semicolon at the end of the DOTPLOT command to produce the prompt SUBC, and the use of the full stop at the end of the final sub-command.

Table 3.9 Sex and type of degree for 40 students (data from Appendix A)

	Type of degree	
Sex	*BA*	*BSc*
Male	2	11
Female	7	20

Sex	Type of degree BA	BSc	
Male	♀ (partial)	♀♀	Key:
Female	♀ ○	♀♀♀ (partial)	♀ = 6 students

Fig. 3.9 Pictogram for the data in Table 3.9.

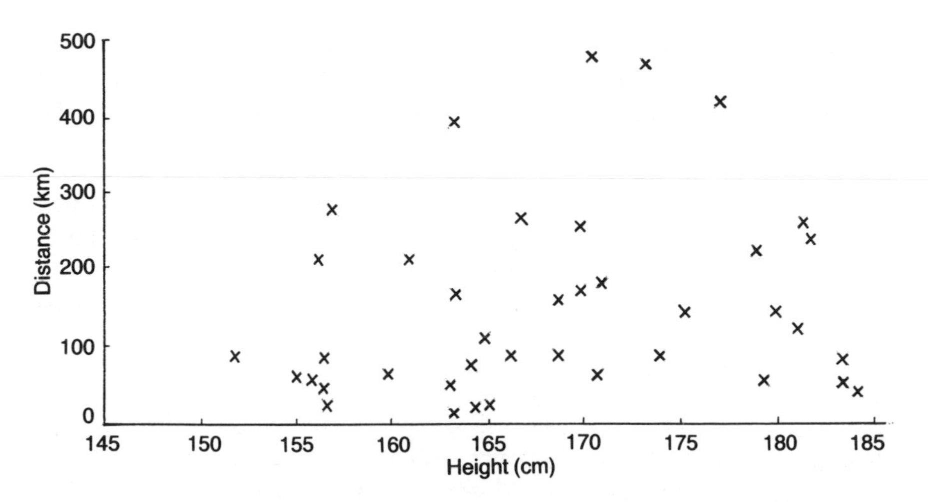

Fig. 3.10 Scatter diagram of the heights and distances from home for 40 students.

Table 3.10 Minitab commands to obtain a scatter diagram like Fig. 3.10

```
MTB> READ C1 C2
DATA> 183 80
DATA> 163  3
.
.
.
.
.
.
.
.
.
.
DATA> 179 45
DATA> END
MTB> NAME C1 'HEIGHT'
MTB> NAME C2 'DISTANCE'
MTB> PLOT C2   C1
```

If we had just written a command DOTPLOT C1 followed by another command DOTPLOT C2, the scales may not have been the same, making comparisons more difficult. Figure 3.11 shows that male heights are on the whole greater than female heights, or does it? The interesting question, 'Is this just a chance difference or a real difference?', cannot be answercd cxccpt subjcctively at this stage. A more objective approach will be described in Chapters 9 and 10.

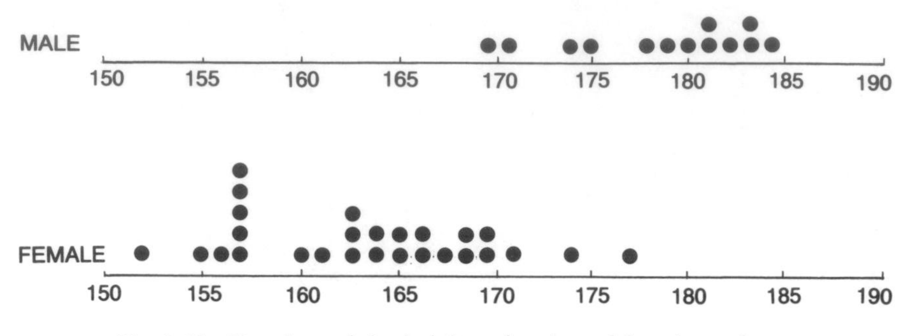

Fig. 3.11 Dotplots of the heights of male and female students.

Table 3.11 Minitab commands to produce dotplots of the heights of male and female students

```
MTB> READ C1 C2
DATA> 1 183
DATA> 2 163
:
.
DATA> 1 179
DATA> END
MTB> NAME C1 'SEX'
MTB> NAME C2 'HEIGHT'
MTB> DOTPLOT C2;
SUBC> BY C1.
```

3.5 SUMMARY

When one-variable or two-variable data are collected for a number of individuals or subjects, these data may be summarized in tables or graphically. Some form of grouping may be advisable if there are many observations; the particular type of table and graph used to summarize the data depends on the type(s) of variable(s). Examples discussed in this chapter are shown in Table 3.12.

WORKSHEET 3: SUMMARIZING DATA BY TABLES AND BY GRAPHICAL METHODS

1. Decide which type of table and graphical method you would use on the following one-variable data sets:
   (a) The number of hours of operation of 100 lightbulbs.
   (b) The type of occupation of 50 adult males.

Table 3.12 Types of table and graph used to summarize data

Number of variables	*Variable type*	*Type of table (and reference)*	*Type of graph (and reference)*
One	Continuous	Ungrouped (Table 3.1)	Dotplot (Fig. 3.1)
			Stem and leaf display (Fig. 3.2)
			Box and whisker plot (Fig. 3.3)
		Grouped frequency (Table 3.2)	Histogram (Fig. 3.4)
		Cumulative frequency (Table 3.3)	Cumulative frequency polygon (Fig. 3.5)
	Discrete	Ungrouped (Table 3.5)	Dotplot (Fig. 3.6)
		Grouped frequency (Table 3.6)	Line chart (Fig. 3.7)
	Categorical	Grouped frequency (Table 3.7)	Bar chart (Fig. 3.8)
Two	Both categorical	Contingency table (Table 3.9)	Pictogram (Fig. 3.9)
	Both continuous	Two columns (Appendix A)	Scatter diagram (Fig. 3.10)
	One continuous, one categorical or discrete	Two columns (Appendix A)	Dotplots (Fig. 3.11)

(c) The total number of earthquakes recorded in the twentieth century for each of ten European coutries.
(d) The percentage of ammonia converted to nitric acid in each of 50 repetitions of an experiment.
(e) The number of hours of operation in a given month for 49 nominally identical computers.
(f) The number of right turns made by 100 rats, each rat having 10 encounters with T-junctions in a maze.
(g) The systolic blood pressure of 80 expectant mothers.
(h) The number of errors (assume a maximum of 5) found by a firm of auditors in 100 balance sheets.
(i) The number of each of six types of room found in a large hotel. The types are: single bedded, double bedded, single and double bedded, each with or without bath.
(j) The density of ten blocks of carbon dioxide.
(k) The number of sheep farms on each type of land. The land types are: flat, hilly and mountainous.
(l) The fluoride content of the public water supply for 100 cities in the UK.

2. The amounts of coffee in grams by which 70 jars of coffee exceeded the nominal 200 g were as follows:

0.7	1.3	1.4	2.2	1.6	0.8	1.2	3.2	2.3	4.6
1.9	1.7	0.2	2.0	2.3	3.1	0.6	2.7	2.9	2.8
1.1	0.7	2.8	1.3	0.3	1.6	3.3	0.4	0.6	5.7
2.3	1.3	2.1	0.9	1.5	2.1	0.9	1.8	3.5	3.5
0.5	2.8	1.6	2.2	0.9	1.2	3.7	1.8	2.0	4.0
1.4	2.7	1.6	2.2	1.1	1.7	1.3	3.4	1.7	3.1
3.0	1.6	0.7	1.8	2.9	1.7	2.2	1.3	2.5	2.7

(a) Draw a dotplot, and comment on the resulting distribution. Summarize the data in a grouped frequency table and draw a histogram. Comment on its shape. Write a Minitab program to read in the above data, obtain a dotplot, stem and leaf display and a (Minitab) histogram. Which of these three graphs do you like the most for these data?
(b) Suppose that the first 35 observations above were obtained on the day shift, while the other 35 were obtained on the night shift. Use two dotplots to compare the shifts, using hand or Minitab methods (please yourself!). Comment on your graphs.

3. For the 'Distance' data in column 6 of Appendix A, draw a histogram using groups 0–49.9, 50–99.9, and so on. Now draw up a cumulative frequency table and polygon. If half the students live less than X km

from Oxford, what is X?. Compare your answer with that obtained from the 'raw' data in Appendix A.

4. Present graphs to help answer the question 'Is the A-level count of Science students more or less the same as the A-level count of Arts students?'.

Summarizing data by numerical measures

Let us condense our
calculations into a few
general abstracts . . .

You are probably familiar with the idea of an 'average' and you may have heard the term 'standard deviation'. Average and standard deviation are examples of numerical measures we use to summarize data. There are many other such measures. It is the purpose of this chapter to show how we calculate some of these measures, but it is equally important for you to learn when to use a particular measure in a given situation. Minitab examples will also be given.

4.1 AVERAGES

In this book the word **average** will be thought of as a vague word meaning 'a middle value' or better 'a single value which in some way represents all the data'. It will only take a definite meaning if we decide that we are referring to a rigorously defined measure such as the

(a) sample (arithmetic) mean, or
(b) sample median, or
(c) sample mode.

Averages will be discussed in sections 4.2 to 4.5 inclusive.

4.2 SAMPLE MEAN ($\bar{x}$)

The sample arithmetic mean, which we will refer to simply as the **sample mean**, of a variable x, is defined in words as follows:

$$\text{sample mean of } x = \frac{\text{sum of the observed values of } x}{\text{number of observed values}}$$

The symbol we use for the sample mean is $\bar{x}$, and its definition in symbols is as follows:

$$\bar{x} = \frac{\Sigma x}{n}$$

where Σx means 'sum the x observations', and n is the number of observations in the sample (refer to section 2.1 if necessary for more detail).

Example

The heights of a sample of 40 students are listed in Appendix A. The sample mean height is

$$\begin{aligned}\bar{x} &= \frac{183 + 163 + \cdots + 184 + 179}{40} \\ &= \frac{6730}{40} \\ &= 168.3\,\text{cm}\end{aligned}$$

The sample mean height of the 40 students is 168.3 cm. Note that we have used one more significant figure than for the raw data.

The formula $\bar{x} = \frac{\Sigma x}{n}$ can be used for both continuous and discrete data, but not for categorical data since the term 'sample mean sex', for example, has no meaning.

The sample mean (and several other measures) can be obtained from Minitab using the command DESCRIBE. For example, if we put the 40 height observations in C1, the command DESCRIBE C1 will produce a mean of 168.25 cm.

4.3 SAMPLE MEDIAN

The **sample median** of a variable x is defined as the middle value when the sample observations of x are ranked in increasing order of magnitude. If there are n observations in a sample, the median is the $\frac{1}{2}(n + 1)$th value.

Example: n *odd*

The heights of five students are 183, 163, 152, 157 and 157 cm.
In rank order: 152, 157, 157, 163, 183.
Here $n = 5$, $(n + 1)/2 = 3$, so the median height is the third value, i.e. 157 cm.

Example: n *even*

The heights of four students are 165, 173, 180 and 164 cm.
In rank order: 164, 165, 173, 180.
Here $n = 4$, $(n + 1)/2 = 2.5$, so the median height is the mean of the second and third values, i.e. $(165 + 173)/2 = 169$ cm.

Example

The heights of a sample of 40 students are listed in column 3 of Appendix A. Instead of ranking these observations we can use one of the following methods:

(a) draw a dotplot (Fig. 3.1);
(b) draw a stem and leaf display (Fig. 3.2);
(c) draw a cumulative frequency polygon (Fig. 3.5);
(d) use Minitab.

We will use all four methods, noting that $n = 40$, hence $(n + 1)/2 = 20.5$, so we need the mean of the 20th and 21st values, assuming that the data are in rank order.

(a) Figure 3.1 shows that the 20th observation is 167, the 21st is 168, so median = $(167 + 168)/2 = 167.5$ cm.
(b) Figure 3.2 shows exactly the same as Fig. 3.1.
(c) Since $40/2 = 20$, we draw a horizontal line on the cumulative frequency polygon at a frequency of 20. Where this line meets the polygon, the corresponding height is the median height (see Fig. 4.1 which gives a median of about 167 cm).
(d) Use the command DESCRIBE as in section 4.2, giving a median of 167.5 cm.

We note that all four methods give the same answer. My personal preference would be to feed the data into a computer and use Minitab. If no computer is available, my preference is for the dotplot.

4.4 SAMPLE MODE

The **sample mode** of a variable x is defined as the observed value which occurs with the highest frequency.

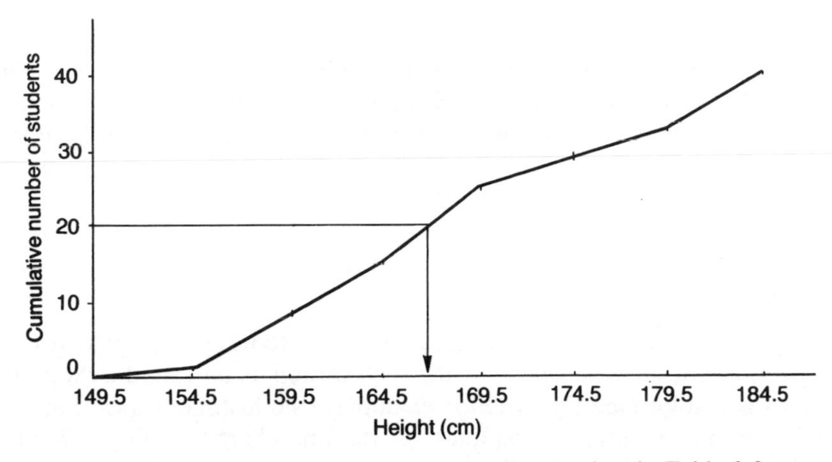

Fig. 4.1 Cumulative frequency polygon for the data in Table 3.3.

Example

The heights of five students are 183, 163, 152, 157, 157 cm.

The mode is 157 cm because it occurs twice, while the others occur only once.

Example

The heights of four students are 165, 173, 180 and 164 cm.

Since each observation occurs the same number of times, we can conclude that either

(a) there is no mode, or
(b) there are four modes.

The fact that the mode may not be unique is one of its disadvantages.

Example

Given the heights of a sample of 40 students in column 3 of Appendix A, we can use either the dotplot (Fig. 3.1) or the stem and leaf display (Fig. 3.2), to obtain a mode of 157 cm, which occurs five times. However, this is hardly a 'middle value'. The modal group, as opposed to the modal height, is 164.5 to 169.5 (see Table 3.2 or Fig. 3.4) and this is perhaps a more useful idea. Note that Minitab does not give the mode when the command DESCRIBE is used.

Example

For categorical data, we cannot calculate either the mean or the median. The mode, on the other hand, may have some limited use. For example if, in our sample of 40 students, 13 are male and 27 are female, then the modal sex is female.

4.5 WHEN TO USE THE MEAN, MEDIAN AND MODE

In order to decide which of the three 'averages' to use in a particular case we need to consider the shape of the distribution as indicated by a graph such as a dotplot (see Fig. 3.1, for example), the histogram (see Fig. 3.4 for a continuous variable example), or the line chart (see Fig. 3.7 for a discrete variable example). For categorical data, the mode is the only average which is defined.

If the shape of the distribution is roughly symmetrical about a vertical centre line, then the sample mean is the preferred average. Such is the case in Figs 3.1 and 3.4, which are graphical plots for the heights of a sample of 40 students. You may have noticed that the mean and median for these data were almost identical, while the mode was not at all representative of the data:

sample mean = 168.2 cm
sample median = 167.5 cm
sample mode = 157 cm

So why should the mean be preferred to the median in this case? The answer is a theoretical one, which you are asked to take on trust, that the sample mean is a more precise measurement for such distributions. ('Precise' here means that, if lots of random samples, each of 40 students, were drawn from the same population of heights, there would be less variation in the sample means than in the sample medians or sample modes of these samples. The terms 'random sample' and 'population' will be discussed in Chapter 8.)

If the shape of the distribution is not symmetrical it is described as **skew**. Figure 4.2 shows sketches of the 'shape' of distributions exhibiting symmetry, positive and negative skewness. It also indicates the rankings of the sample mean, sample median and sample mode in each of the three cases. For markedly skew data there will be a small number of extremely high values (Fig. 4.2(b)) or low values (Fig. 4.2(c)) which are not balanced by values on the other side of the distribution. The sample mean is more influenced by these few extreme values than the sample median. So the sample median is preferred for data showing marked skewness. By 'marked skewness' we mean that the measure of skewness

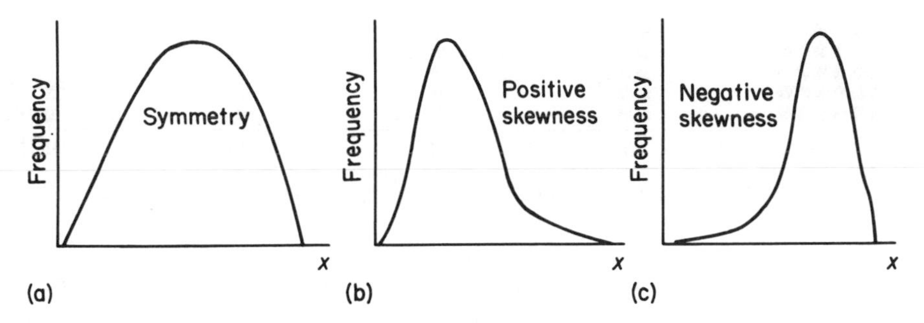

Fig. 4.2 Symmetry and skewness: (a) mean = median = mode; (b) mean > median > mode; (c) mean < median < mode.

(see section 4.12) is greater than 1, or less than −1, as a rough guide. If in doubt both the sample mean and sample median should be quoted.

The mode is not of much use for either continuous or discrete data, since it may not be unique (as we have seen in section 4.4), it may not exist at all, and for other theoretical reasons. The mode is useful only for categorical data.

Occasionally, distributions arise for which none of the three 'averages' is particularly informative.

Example

Table 4.1 shows the number of cigarettes smoked by 50 subjects. Drawing a dotplot (see Fig. 4.3) shows a positively skew distribution. The mean number of cigarettes smoked per day is = $(0 \times 30 + 10 \times 10 + 20 \times 5 + 30 \times 3 + 4 \times 2)/50 = 7.4$. However, this number does not seem to 'represent' the data very well. The median number smoked is zero, since the 25th and the 26th values are both zero. The mode is also zero. However, zero does not seem to represent the data very well either.

Table 4.1 and Fig. 4.3 are both very informative, but if you must

Table 4.1 The number of cigarettes smoked per day by 50 subjects

Number of cigarettes	*Number of subjects*
0	30
10	10
20	5
30	3
40	2

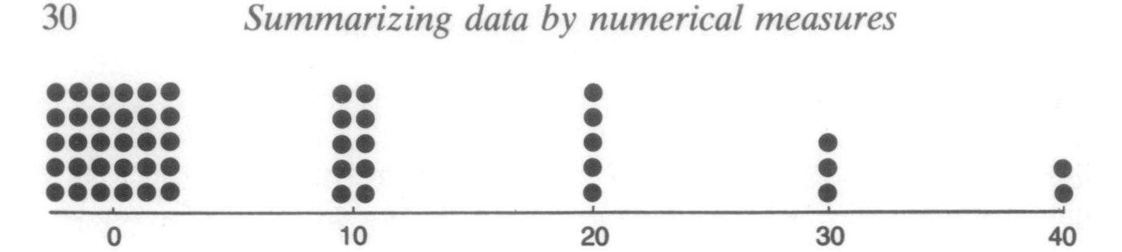

Fig. 4.3 Dotplot of the number of cigarettes smoked per day by 50 subjects.

summarize these data numerically, you could state that '60% of subjects are non-smokers, while the mean number of cigarettes per day for smokers is 18.5'.

4.6 MEASURES OF VARIATION

Averages are not the whole story. They do not give a complete description of a set of data and can, on their own, be misleading. So the definition of a statistician as one who, on plunging one foot into a bucket of boiling water and the other into a bucket of melting ice, declares 'On average I feel just right!' completely misses the purpose of statistics, which is to collect and analyse data which vary. It is not the aim of this book to lament the misconceptions some people have about statistics, but hopefully to inform and educate. So it would be more reasonable for the caricatured statistician to feel unhappy because the temperature of his feet varies so greatly about a comfortable average. Similarly, an employee feels unhappy when told 'wages have risen by 10% in the past year', if his own wage has risen by only 3%, while the cost of living has risen by 8% (both the 8% and the 10% are averages, by the way).

Two measures of variation will be discussed in sections 4.7, 4.8 and 4.9 in some detail, and three other measures of variation will be mentioned briefly in section 4.11.

4.7 SAMPLE STANDARD DEVIATION (s)

One way of measuring variation in sample data is to sum the differences between each observed value and the sample mean, x, to give:

$$\Sigma(x - \bar{x}).$$

However, this always gives the answer zero, as we saw in section 2.1 and Worksheet 2, Question 4.

A more useful measure of variation, called the **sample standard deviation**, s, is obtained by summing the squares of the differences $(x - \bar{x})$, dividing by $n - 1$ (where n is the number of sample observations), and then taking the square root. This gives a kind of 'root mean square deviation' (see the formula for s below).

The reason for squaring the differences is that this makes them all positive or zero. The reason for dividing by $n - 1$ rather than n is discussed later in this section. The reason for taking the square root is to make the measure s have the same units as the variable x. There are more theoretical reasons than these for using standard deviation as a measure of variation, but I hope the above will give you an intuitive feel for the formulae which are now introduced.

The sample standard deviation, s, may be defined by the formula:

$$s = \sqrt{\frac{\Sigma (x - \bar{x})^2}{n - 1}}$$

An alternative form of this formula which is easier to use for calculation purposes is:

$$s = \sqrt{\frac{\Sigma x^2 - \frac{(\Sigma x)^2}{n}}{n - 1}}$$

Example

The heights (cm) of a sample of five people are 183, 163, 152, 157 and 157. Therefore:

$$\begin{aligned}\Sigma x &= 183 + 163 + 152 + 157 + 157 = 812 \\ \Sigma x^2 &= 183^2 + 163^2 + 152^2 + 157^2 + 157^2 = 132\,460 \\ n &= 5.\end{aligned}$$

The sample standard deviation is

$$\begin{aligned}s &= \sqrt{\frac{\Sigma x^2 - \frac{(\Sigma x)^2}{n}}{n - 1}} \\ &= \sqrt{\frac{132\,460 - \frac{812^2}{5}}{4}} \\ &= 12.2\,\text{cm}\end{aligned}$$

Example

For the heights of a sample of 40 students given in column 3 of Appendix A, we have as follows:

$$\Sigma x = 6730, \quad \Sigma x^2 = 1\,135\,558, \quad n = 40$$

so we calculate the sample standard deviation as follows:

$$s = \sqrt{\frac{1\,135\,558 - \frac{6730^2}{40}}{39}}$$

$$= 9.1\,\text{cm}$$

Notes

(a) The units of standard deviation are the same as the units of the variable height, i.e. centimetres, in both examples above.
(b) The answer should be given to one more significant figure than the raw data, i.e. one decimal place in both the examples above.

A question which is often asked is, 'Why use $n - 1$ in the formulae for s?'. The answer is that the values we obtain give better estimates of the standard deviation of the population than would be obtained if we used n instead. In what is called 'statistical inference' (Chapter 8 onwards) we are interested not so much in sample data, as in what conclusions, based on sample data, can be drawn about the population from which the sample was taken.

Another natural question at this stage is, 'Now that we have calculated the sample standard deviation, what does it tell us?'. The answer is 'Be patient!'. When we have discussed the normal distribution in Chapter 7, standard deviation will become more meaningful. For the moment please accept the basic idea that standard deviation is a measure of variation about the mean. The more variation in the data, the higher will be the standard deviation. If there is no variation at all, the standard deviation will be zero. It can never be negative.

To obtain the sample standard deviation using Minitab, use the command DESCRIBE C1, assuming the relevant data are stored in C1.

4.8 SAMPLE INTER-QUARTILE RANGE

Just as the sample median is such that half the sample observations are less than it, and it is the $((n + 1)/2)$th value, we define the lower and upper quartiles similarly.

The lower quartile is such that one-quarter of the sample observations are less than it, and it is the $((n + 1)/4)$th value.

The upper quartile is such that three-quarters of the sample observations are less than it, and it is the $3((n + 1)/4)$th value.

The **sample inter-quartile range** is defined as the difference between the quartiles, that is,

$$\text{sample inter-quartile range} = \text{upper quartile} - \text{lower quartile}.$$

Example

The heights (cm) of a sample of five people are 183, 163, 152, 157 and 157.

In rank order these are 152 157 157 163 183.

Since $n = 5$,

$$(n+1)/4 = 1\tfrac{1}{4}, \text{ lower quartile} = 152 + \tfrac{1}{4}(157 - 152) = 153.2,$$
$$3(n+1)/4 = 3\tfrac{3}{4}, \text{ upper quartile} = 157 + \tfrac{3}{4}(163 - 157) = 161.5$$
$$\text{inter-quartile range} = 161.5 - 153.2 = 8.3\,\text{cm}.$$

Example

For the heights of a sample of 40 students given in column 3 of Appendix A, instead of ranking the 40 observations, we could use the same four methods for finding the lower and upper quartiles as we used for finding the median (see section 4.3). We refer to only two methods here:

(a) Minitab and the command DESCRIBE C1, assuming the relevant data are in C1. The Minitab output refers to the lower and upper quartiles as Q1 and Q3, respectively. For our height data, Q1 = 161.5, Q3 = 176.5, and sample inter-quartile range = Q3 − Q1 = 176.5 − 161.5 = 15 cm.

(b) Dotplot (see Fig. 3.1). Since $n = 40$, $(n + 1)/4 = 10\tfrac{1}{4}$; the 10th observation is 161, the 11th is 163, so Q1 = $161 + \tfrac{1}{4}(163 - 161) = 161.5$. Also $3(n + 1)/4 = 30\tfrac{3}{4}$; the 30th observation is 175, the 31st is 177,
so Q3 = $175 + \tfrac{3}{4}(177 - 175) = 176.5$, and (finally!),
sample inter-quartile range = 176.5 − 161.5 = 15 cm, as above.

Notice that the 'middle half' of the sample observations lie between the lower and upper quartiles.

4.9 WHEN TO USE STANDARD DEVIATION AND INTER-QUARTILE RANGE

In order to decide which of these two measures of variation to use in a particular case, the same considerations apply as for averages (refer to section 4.5 if necessary). For roughly symmetrical data, use standard deviation. For markedly skew data, use inter-quartile range.

4.10 BOX AND WHISKER PLOTS

These plots were mentioned in section 3.2 (see Fig. 3.3) before the terms median, lower quartile and upper quartile had been introduced. Figure 3.3 is reproduced here as Fig. 4.4 for convenience. From left to right

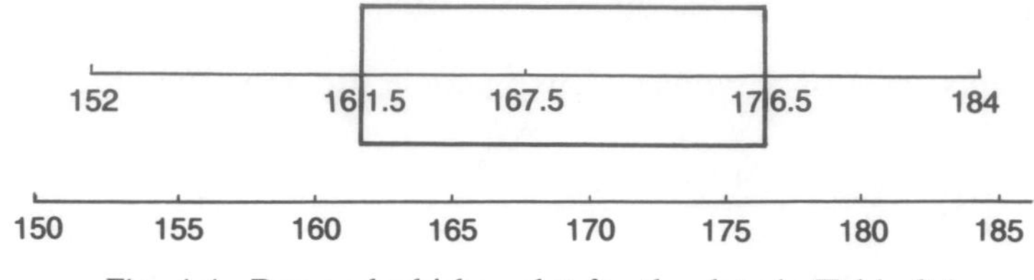

Fig. 4.4 Box and whisker plot for the data in Table 3.1.

the five values 152, 161.5, 167.5, 176.5 and 184 are respectively the minimum (smallest), lower quartile (Q1), median, upper quartile (Q3) and maximum (largest) values in the sample. You can obtain a plot like Fig. 4.4 using the Minitab command BOXPLOT C1 (assuming that the relevant data are held in C1). It is sometimes useful to compare two subsets of data using the subcommand BY in a similar way to the method used for the two dotplots (see Table 3.11 and Fig. 3.11). The five statistics available from Fig. 4.4 can all be obtained by other methods, e.g. the Minitab command DESCRIBE, or, more indirectly, from a dotplot or a stem and leaf display.

4.11 OTHER MEASURES OF VARIATION

We will consider three other measures of variation briefly:

(a) **Variance** is simply the square of the standard deviation, so we can use the symbol s^2. Variance is a common term in statistical methods beyond the scope of this book, for example in 'analysis of variance'.
(b) **Coefficient of variation** is defined as $100s/\bar{x}$, and is expressed as a percentage. This is used to compare the variabilities of two sets of data when there is an obvious difference in magnitude in both the means and standard deviations. For example, to compare the variation in the heights of boys aged 5 and 15 years, suppose $\bar{x}_5 = 100$, $s_5 = 6$, $\bar{x}_{15} = 150$, $s_{15} = 9$, then both sets have a coefficient of variation of 6%.
(c) **Range** is defined as the difference between the largest observed value and the smallest observed value, when we are discussing sample data. It is commonly used because it is simple to calculate, but it is unreliable, except in special circumstances, because only two of the sample observations are used to calculate it. Also the more sample observations we take, the larger the range is likely to be.

4.12 A MEASURE OF SKEWNESS

We saw in section 4.5 that if the distribution of a set of data is perfectly symmetrical, then the mean and median are equal. If there is positive skewness, the mean exceeds the median, while the mean is less than the median for negatively skew data. The following dimensionless measure is therefore zero, positive or negative depending on the type of skewness:

$$\frac{3(\text{sample mean} - \text{sample median})}{\text{sample standard deviation}}$$

As a rough guide, if this measure is greater than 1, we can say that the distribution is markedly positively skew. If it is less than -1 we can conclude marked negative skewness.

Example

For the heights of 40 students given in column 3 of Appendix A,

$$\begin{aligned} \text{sample mean} &= 168.2\,\text{cm} \\ \text{sample median} &= 167.5\,\text{cm} \\ \text{sample standard deviation} &= 9.1\,\text{cm} \\ \text{measure of skewness} &= \frac{3(168.2 - 167.5)}{9.1} = 0.23 \end{aligned}$$

The distribution of heights is very slightly positively skew.

4.13 SUMMARY

When a variable is measured for a number of individuals, the resulting data may be summarized by calculating averages and measures of variation. In addition, a measure of skewness is sometimes useful. The particular type of average and measure of variation required depends on the type of variable and shape of the distribution. Some examples are given in Table 4.2. Three other measures of variation are the variance, coefficient of variation and range.

Table 4.2 Examples of averages and measures of variation

Type of variable	*Shape of distibution*	*Average*	*Measure of variation*
Continuous or discrete	Roughly symmetrical, unimodal	Sample mean ($\bar{x}$)	Sample standard deviation (s)
Continuous or discrete	Markedly skew, unimodal	Sample median	Sample inter-quartile range
Categorical		Sample mode	

WORKSHEET 4: SUMMARIZING DATA BY NUMERICAL MEASURES

1. (a) Why do we need averages?
   (b) Which average can have more than one value?
   (c) Which average represents the value when the total of all the sample observations is shared out equally?
   (d) Which average has the same number of observations above it as below it?
   (e) When is the sample median preferred to the sample mean?
   (f) When is the sample mode preferred to the sample mean?
   (g) When is the sample mean preferred to both the sample median and the sample mode?
2. (a) Why do we need measures of variation?
   (b) What measure of variation is most useful in the case of: (i) a symmetrical distribution, (ii) a skew distribution?
   (c) Think of an example of sample data where the range would be a misleading measure of variation.
   (d) Name the measure of variation associated with the: (i) sample mean, (ii) sample median, (iii) sample mode.
   (e) Name the average associated with the: (i) sample standard deviation, (ii) sample inter-quartile range, (iii) range.
3. The weekly incomes (£) of a random sample of self-employed window cleaners are:

   75, 67, 60, 62, 65, 67, 62, 68, 82, 67, 62, 200.

   (a) Find the sample mean, sample median and sample mode of weekly income. Why are your three answers different?
   (b) Find the sample standard deviation and sample inter-quartile range of weekly income. Why are your answers different?
   (c) Which of the measures you have obtained are the most useful in summarizing the data?

   Try this question by hand calculation, and check your answers using Minitab.
4. Eleven cartons of sugar, each nominally containing 1 kg, yielded the following weights of sugar.

   1.02, 1.05, 1.08, 1.03, 1.00, 1.06, 1.08, 1.01, 1.04, 1.07, 1.00.

   Calculate the sample and sample standard deviation of the weight of sugar.
   Try this question by calculator and by Minitab.

5. Using the data in Question 2 of Worksheet 3, find:
   (a) the sample mean and standard deviation,
   (b) the sample median and inter-quartile range.
   Decide which is the preferred (i) average and (ii) measure of variation.

6. For the distance data in column 6 of Appendix A, find:
   (a) the sample mean and standard deviation,
   (b) the sample median and inter-quartile range.
   Decide which is the preferred (i) average and (ii) measure of variation.

7. For the distance data in column 6 of Appendix A, compare the distances of male and female students, using the Minitab commands DESCRIBE, DOTPLOT and BOXPLOT. (Hint: use the subcommand BY as in Table 3.11.)

Probability

Dr Price estimates the chance in favour of the wife being the survivor in marriage as 3 to 2.

5.1 INTRODUCTION

The opening chapters of this book have been concerned with statistical data and methods of summarizing such data. We can think of such sample data as having been drawn from a larger 'parent' population. Conclusions from sample data about populations (which is a branch of statistics called 'statistical inference', discussed from Chapter 8 onwards) must necessarily be subject to some uncertainty since the sample cannot contain all the information in the population. This is one of the main reasons why probability, which is a measure of uncertainty, is now discussed.

Probability is a topic which may worry you, either if you have never studied it before, or if you have studied it before but you did not fully get to grips with it. It is true that the study of probability requires a clear head, a logical approach and the ability to list all the outcomes of simple experiments, often with the aid of diagrams. After some experience and some (possibly painful) mistakes, which are all part of the learning process, the penny usually begins to drop.

Think about the following question which will give you some feedback on your present attitude towards probability (try not to read the discussion until you have thought of an answer).

Example 5.1

A person tosses a coin five times. Each time it comes down heads. What is the probability that it will come down heads on the sixth toss?

Discussion
If your answer is '1/2' you are assuming that the coin is 'fair', meaning that it is equally likely to come down heads or tails. You have ignored the 'data' that all five tosses resulted in heads.

If your answer is 'less than 1/2' you may be quoting 'the law of averages' which presumably implies that, in the long run, half the tosses will result in heads and half in tails. This again assumes that the coin is fair. Also, do six tosses constitute a long run of tosses, and does the 'law of averages' apply to each individual toss?

If your answer is 'greater than 1/2', perhaps you suspect that the coin has two heads, in which case the probability of heads would be 1, or that the coin has a bias in favour of heads. Think about this teasing question again when you have read this chapter.

5.2 BASIC IDEAS OF PROBABILITY

One dictionary definition of probability is 'the extent to which an event is likely to occur, measured by the ratio of the favourable cases to the whole number of cases possible'. Consider the following example.

Example 5.2

A ball is selected at random from a bag containing three red balls and seven white balls (Fig. 5.1). The probability that a red ball will be drawn is 3/10. Note the following points:

(a) 'At random' means that each of the ten balls has the same chance (probability) of being selected, implying that we mix up the balls and the person selecting the ball should look away or close his/her eyes. We say that the 10 outcomes are 'equally likely' in this case.
(b) Probability is a measure of uncertainty which, as we shall see later, can take any value between 0 and 1.
(c) The probability that a white ball is drawn is 7/10. Note that the total of the two probabilities is $3/10 + 7/10 = 1$, and that no other outcome is possible.

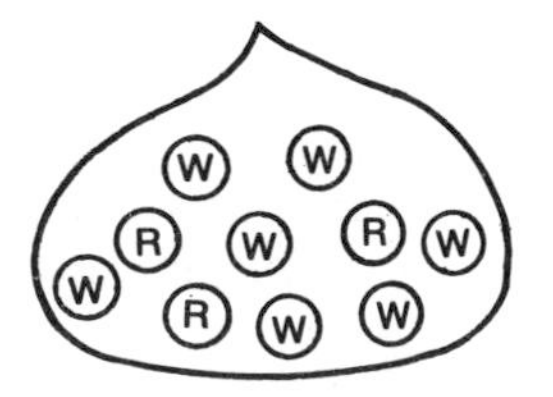

Fig. 5.1 A bag containing three red and seven white balls.

Does Fig. 5.1 help you to understand Example 5.2? It helps me (!) to visualize a probability problem either in my head or on paper, and the more complex the problem the more useful a visual aid is likely to be, as we shall see later in this chapter.

Recalling the dictionary definition of probability above, the number of favourable cases is 3 for the event 'red ball', out of a total of 10 possible cases, and the required probability is again 3/10.

In order to gain an understanding of probability, it is helpful to define three terms which have a special meaning when we discuss probability.

A **trial** is an action which results in one of several possible outcomes.
An **experiment** is a series of trials (or possibly just one).
An **event** is a set of outcomes with something in common.

In Example 5.2 above, the trial is 'drawing a ball from a bag', the experiment is also 'drawing a ball from a bag', since only one ball is selected, the event is 'red ball', corresponding to three of the ten possible outcomes.

5.3 THE *A PRIORI* DEFINITION OF PROBABILITY FOR EQUALLY LIKELY OUTCOMES

This section is a more formal look at a definition of probability for experiments whose outcomes are equally likely, as in Example 5.2.

Suppose each trial in an experiment can result in one of n 'equally likely' outcomes, r of which correspond to an event, E. Then the probability of event E is r/n, which we write

$$P(E) = \frac{r}{n}$$

This *a priori* definition has been used for Example 5.2; event E is 'red ball', $n = 10$ since it is assumed that each of the ten balls is equally likely to be drawn from the bag, and $r = 3$ since three of the ten balls are red and therefore correspond to the event E. So we write

$$P(\text{red ball}) = \frac{3}{10}$$

Note the following points:

(a) We only have to think about the possible outcomes, we do not actually have to carry out an experiment of removing balls from a bag. The Latin phrase *a priori* means 'without investigation or sensory experience'.
(b) It is necessary to know that the possible outcomes are equally likely to occur. This is why this definition is called a 'circular' definition,

since equally likely and equally probable have the same meaning. More importantly we should not use the *a priori* definition if we do not know that the possible outcomes are equally likely. (Example: 'Either I will become the manager of the England soccer team or I will not, so the probability that I will is 1/2, and the same probability applies to everybody'. This is clearly an absurd argument.)

The *a priori* definition is most useful in games of chance, i.e. gambling.

5.4 THE RELATIVE FREQUENCY DEFINITION OF PROBABILITY, BASED ON EXPERIMENTAL DATA

Example 5.3

Suppose we consider all the couples who married in 1990 in the UK. What is the probability that, if we select one such couple at random, they will still be married to each other today? We have no reason to assume that the outcomes 'still married' and 'not still married' are equally likely. We need to look at the experimental data, that is to carry out a survey to find out what state those marriages are in today, and the required probability would be the ratio, i.e. the relative frequency, given by:

$$\frac{\text{number of couples still married today who married in 1990 in the UK}}{\text{number of couples who married in 1990 in the UK}}$$

We will not consider the practical problems of gathering such data!

Formally, the relative frequency definition of probability is as follows. If in a large number of trials, n, r of these trials result in event E, the probability of event E is r/n.

So we write

$$P(E) = \frac{r}{n}$$

Notes

(a) The number of trials, n, must be large. The larger the value of n, the better is the estimate of probability.
(b) The phrase 'at random' above means that each couple has the same chance of being selected from the particular population.
(c) One theoretical problem with this definition of probability is that apparently there is no guarantee that the value of r/n will settle down to a constant value as the number of trials gets larger and larger. Luckily this is not a practical problem, so we shall not pursue it.

5.5 THE RANGE OF POSSIBLE VALUES FOR A PROBABILITY VALUE

Using either of the two definitions of probability, we can show that probabilities can only take values between 0 and 1. The value of r must take one of the integer values between 0 and n, so r/n can take values between $0/n$ and n/n, that is 0 and 1.

If $r = 0$ we are thinking of an event which cannot occur (*a priori* definition) or an event which has not occurred in a large number of trials (relative frequency definition). For example, the probability that we will throw a 7 with an ordinary die is 0.

If $r = n$ we are thinking of an event which must always occur (*a priori* definition) or an event which has occurred in every one of a large number of trials (relative frequency definition). For example, the probability that the sun will rise tomorrow can be assumed to be 1, unless you are a pessimist (see section 5.7).

5.6 PROBABILITY, PERCENTAGE, PROPORTION AND ODDS

We can convert a probability to a percentage by multiplying it by 100. So a probability of 3/4 implies a percentage of 75%.

We can also think of probability as meaning the same thing as proportion. So a probability of 3/4 implies that the proportion of times an event will occur is also 3/4.

A probability of 3/4 is equivalent to odds of 3/4 to 1/4, which is usually expressed as 3 to 1.

You are advised to try Worksheet 5, Questions 1 to 7, before proceeding with this chapter.

5.7 SUBJECTIVE PROBABILITY

There are other definitions of probability apart from the two discussed earlier in this chapter. We all use **subjective probability** in forecasting future events, for example when we try to decide whether it will rain tomorrow, and when we try to assess the reactions of others to our opinions and actions. We may not be quite so calculating as to estimate a probability value, but we may regard future events as being probable, rather than just possible. In subjective assessments of probability we may take into account experimental data from past events, but we are likely to add a dose of subjectivity depending on our personality, our mood and other factors.

5.8 PROBABILITIES INVOLVING MORE THAN ONE EVENT

Suppose that we are interested in the probabilities of two possible events, E_1 and E_2. For example, we may wish to know the probability that both events will occur, or the probability that either or both events will occur. We will refer to these as, respectively:

$$P(E_1 \text{ and } E_2) \quad \text{and} \quad P(E_1 \text{ or } E_2 \text{ or both}).$$

In set theory notation these compound events are called the **intersection** and **union** of events E_1 and E_2 respectively, and their probabilities are written:

$$P(E_1 \cap E_2) \quad \text{and} \quad P(E_1 \cup E_2).$$

There are two probability laws which can be used to estimate such probabilities, and these are discussed in sections 5.9 and 5.10.

5.9 MULTIPLICATION LAW (THE 'AND' LAW)

The general case of the **multiplication law** is:

$$P(E_1 \text{ and } E_2) = P(E_1)\,P(E_2 \mid E_1)$$

where $P(E_2 \mid E_1)$ means the probability that event E_2 will occur, given that event E_1 has already occurred. The vertical line between E_2 and E_1 should be read as 'given that' or 'on the condition that'. $P(E_2 \mid E_1)$ is an example of what is called a **conditional** probability.

Example 5.4

If two cards are selected at random, one at a time without replacement from a pack of 52 playing cards, what is the probability that both cards will be aces?

$$\begin{aligned} \text{P(two aces)} &= \text{P(first card is an ace and second card is an ace)} \\ &= \text{P(first card is an ace)} \times \text{P(second card is an ace} \mid \\ &\quad \text{first card is an ace)} \end{aligned}$$

Using the multiplication law, where E_1 is the event that the first card is an ace, and E_2 is the event that the second card is an ace,

$$\begin{aligned} \text{P(two aces)} &= \frac{4}{52} \times \frac{3}{51} \quad \text{(see Fig. 5.2)} \\ &= 0.0045 \end{aligned}$$

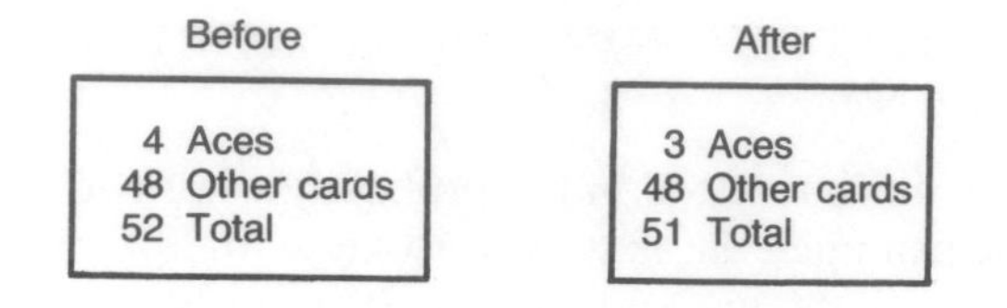

Fig. 5.2 Before and after the first card is drawn, without replacement.

(Note that 4 decimal places are usually more than sufficient for a probability value.)

In many practical examples the probability of event E_2 does not depend on whether E_1 has occurred. In this case we say that events E_1 and E_2 are **statistically independent**, giving rise to the special case of the multiplication law for independent events E_1 and E_2:

$$P(E_1 \text{ and } E_2) = P(E_1) \times P(E_2)$$

Example 5.5

Consider Example 5.4, and change the phrase 'without replacement' to 'with replacement'.

P(two aces) = P(first card is an ace and second card is an ace), as in Example 5.4,
= P(first card is an ace) × P(second card is an ace)

Using the multiplication law for independent events,

$$\text{P(two aces)} = \frac{4}{52} \times \frac{4}{52} \quad \text{(see Fig. 5.3)}$$
$$= 0.0059$$

Because the first card is replaced, the probability that the second card is an ace will be 4/52 whatever the first card is.

Note the following points:

(a) Comparing the two cases of the multiplication law we can state that, if two events E_1 and E_2 are statistically independent,

$$P(E_2 | E_1) = P(E_2)$$

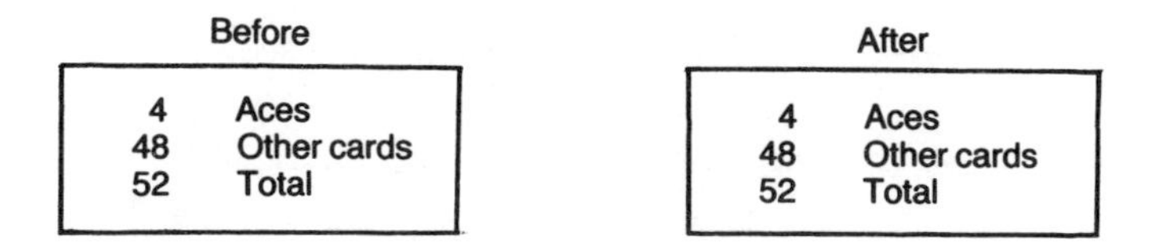

Fig. 5.3 Before and after the first card is drawn, with replacement.

(b) Clearly we could write the general law of multiplication alternatively as:

$$P(E_1 \text{ and } E_2) = P(E_2) \times P(E_1 \mid E_2)$$

by swapping E_1 and E_2.

5.10 ADDITION LAW (THE 'OR' LAW)

The general case of the **addition law** is:

$$P(E_1 \text{ or } E_2 \text{ or both}) = P(E_1) + P(E_2) - P(E_1 \text{ and } E_2)$$

Example 5.6

If a die is tossed twice, what is the probability of getting at least one 5?

Here, defining E_1 as the event '5 on first toss', and E_2 as the event '5 on second toss',

$$\begin{aligned} \text{P(at least one 5)} &= \text{P(5 on first toss or 5 on second toss or 5 on both tosses)} \\ &= \text{P(5 on first toss)} + \text{P(5 on second toss)} - \text{P(5 on both tosses)} \end{aligned}$$

using the addition law,

$$= \frac{1}{6} + \frac{1}{6} - \frac{1}{6} \times \frac{1}{6}$$

using the multiplication law for independent events for P(5 on both tosses) (see Fig. 5.4)

$$= \frac{6}{36} + \frac{6}{36} - \frac{1}{36}$$

$$= \frac{11}{36}$$

$$= 0.3056$$

1st toss outcomes

① ② ③
④ ⑤ ⑥

2nd toss outcomes

① ② ③
④ ⑤ ⑥

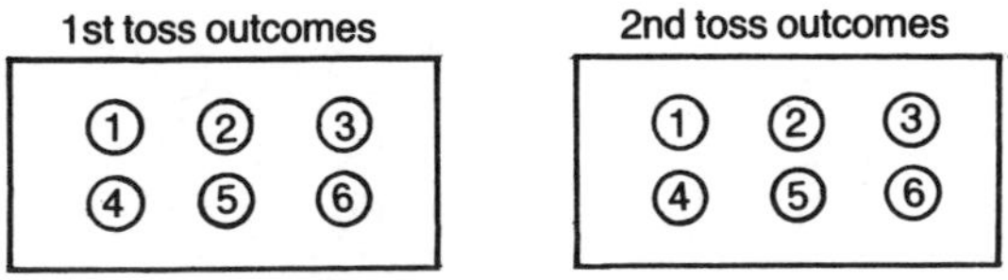

Fig. 5.4 Tossing a die twice.

In many practical cases the events E_1 and E_2 are such that they cannot both occur. In this case we say that the events E_1 and E_2 are **mutually exclusive**, giving rise to the special case of the addition law for mutually exclusive events E_1 and E_2:

$$P(E_1 \text{ or } E_2) = P(E_1) + P(E_2)$$

If you compare the versions of the addition law, you will see that if E_1 and E_2 are mutually exclusive,

$$P(E_1 \text{ and } E_2) = 0$$

Example 5.7

If a die is tossed twice, what is the probability that the total score is 11?

There are only two ways of getting a total of 11, either by getting 6 on the first toss and 5 on the second toss, or by getting 5 on the first toss and 6 on the second toss. These two ways, which we will call E_1 and E_2, are mutually exclusive.

$$\begin{aligned} P(\text{total of } 11) &= P(E_1) + P(E_2) \\ &= \frac{1}{6} \times \frac{1}{6} + \frac{1}{6} \times \frac{1}{6} \\ &= 0.0556 \end{aligned}$$

We could have obtained the answer to Example 5.7 by noting, from Fig. 5.5, that 2 of the 36 equally likely totals (each has a probability of 1/36) result in a total of 11. Hence, using the *a priori* definition of probability,

$$P(11) = 2/36 = 0.0556$$

Similarly, the answer to Example 5.6 could be obtained by using the display shown in Fig. 5.6. Observe that P(5 on first toss or 5 on second toss or both) = 11/36 = 0.3056, as before, since there are 11 favourable cases out of 36, and all 36 are equally likely.

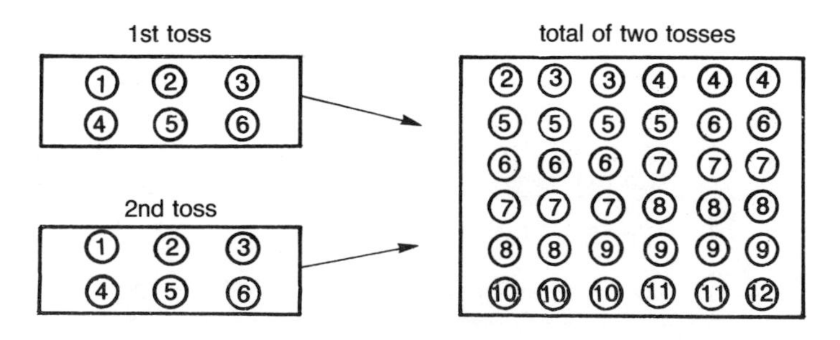

Fig. 5.5 The total score when a die is tossed twice.

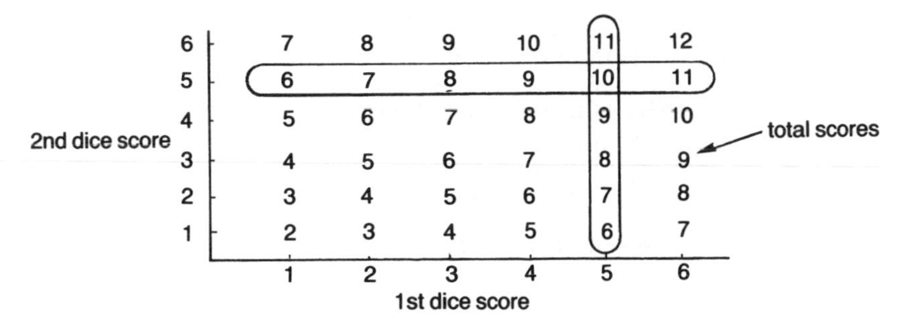

Fig. 5.6 The total score when two dice are thrown.

5.11 MUTUALLY EXCLUSIVE AND EXHAUSTIVE EVENTS

If all possible outcomes of an experiment are formed into a set of mutually exclusive events, we say they form a **mutually exclusive and exhaustive** set of events which we will call $E_1, E_2, \ldots, E_n$, if there are n events. Applying the special case of the law of addition,

$$P(E_1 \text{ or } E_2 \text{ or} \ldots \text{or } E_n) = P(E_1) + P(E_2) + \cdots + P(E_n)$$

But since the events are exhaustive, one of them must occur, and so the left-hand side of the equation is 1. In words, then, this result is:

> The sum of the probabilities of a set of mutually exclusive and exhaustive events is 1.

This result is useful in checking whether we have correctly calculated the separate probabilities of the various mutually exclusive events of an experiment. It is also a result for which we will find an application in Chapter 6.

Example 5.8

For families with two children, what are the probabilities of the various possibilities, assuming that boys and girls are equally likely at each birth?

Four mutually exclusive and exhaustive events are BB, BG, GB and GG, where, for example, BG means a boy followed by a girl. Using the special law of multiplication,

$$P(BG) = \frac{1}{2} \times \frac{1}{2} = \frac{1}{4}$$

Similarly $P(BB) = P(GB) = P(GG) = \frac{1}{4}$, and the total probability is 1.

5.12 COMPLEMENTARY EVENTS AND THE CALCULATION OF P(AT LEAST 1 . . .)

For any event E, there is a **complementary** event E', which we call 'not E'. Since either E or E' must occur, and they cannot both occur,

$$P(E) + P(E') = 1,$$

a special case of the result of the previous section, It follows that:

$$P(E) = 1 - P(E')$$

This result is useful in some calculations where we will find it easier to calculate the probability of the **complement** to some event and subtract the answer from 1 than it is to calculate directly the probability of the event. This is especially true when we wish to calculate the probability that 'at least 1 of something will occur in a number of trials', since:

$$P(\text{at least } 1 \ldots) = 1 - P(\text{none} \ldots).$$

Example 5.9

For families with four children, what is the probability that there will be at least one boy, assuming boys and girls are equal likely?

Instead of listing the 16 outcomes BBBB, BBBG, etc., we simply say:

$$\begin{aligned} P(\text{at least 1 boy}, &= 1 - P(\text{no boys}) \\ &= 1 - P(\text{GGGG}) \\ &= 1 - \frac{1}{2} \times \frac{1}{2} \times \frac{1}{2} \times \frac{1}{2} \\ &= \frac{15}{16} \end{aligned}$$

5.13 PROBABILITY TREES

A **probability tree** (sometimes called a **tree diagram**) can be used instead of the laws of probability when we are considering the outcomes of an experiment consisting of a sequence of a few trials.

In general, in a probability tree, the events resulting from the first trial are represented by two or more branches from a starting point. More branches are added to represent the events resulting from the second and subsequent trials. The branches are labelled with the names of the events and their probabilities, taking into account previous branches (so we might be dealing with conditional probabilities as in Example 5.4). In order to calculate the probabilities of the set of mutually exclusive and exhaustive events, corresponding to all the ways of getting from the starting point to the end of the branch, probabilities are multiplied (so we are really using the law of multiplication). The total probability should, of course, equal 1.

Example 5.10

Two cards are selected at random, one at a time without replacement, from a pack of 52 playing cards. Let A be the event that an ace is selected, and A' the event that the card is not an ace. The results of the experiment are shown in Fig. 5.7. Hence:

P(two aces) = P(AA) = 0.0045
P(one ace) = P(AA' or $A'A$) = P(AA') + P($A'A$) = 0.1448
P(no aces) = P($A'A'$) = 0.8507

In general, the total number of branches needed will depend on both the number of possible events resulting from each trial and the number of trials. For example, if each of four trials can result in one of two possible outcomes there will be 2 + 4 + 8 + 16 = 30 branches, or if each of three trials can result in one of three possible outcomes there will be 3 + 9 + 27 = 39 branches.

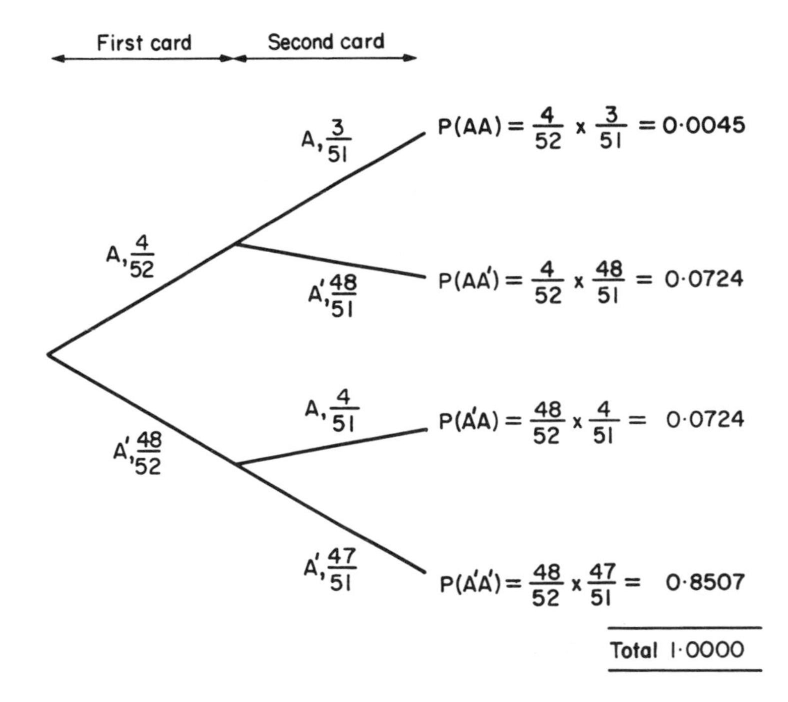

Fig. 5.7 Probability tree for two cards, without replacement.

Clearly it becomes impractical to draw a probability tree if the number of branches is large. In the next chapter we will discuss more powerful ways of dealing with any number of independent trials each with only two possible outcomes.

5.14 VENN DIAGRAMS AND REES DIAGRAMS

Venn diagrams can be used to help us to disentangle probability problems. Strictly speaking, a Venn diagram is used to describe 'sets', represented by areas within a larger area, and mutually exclusive 'elements' within sets, as shown, for example, in Fig. 5.8.

If we take this a stage further and make the area corresponding to an element equal to the probability of that element, and make the total area equal to (the total probability of) 1, we have a modified diagram, which I will call a **Rees diagram** (Fig. 5.9), unless someone else has claimed it already! In both Fig. 5.9(a) and 5.9(b), the total area of the large rectangle equals 1 and this is subdivided into six equal areas of 1/6, the probability of each outcome.

This idea can be used in more complicated examples, as Fig. 5.10 shows.

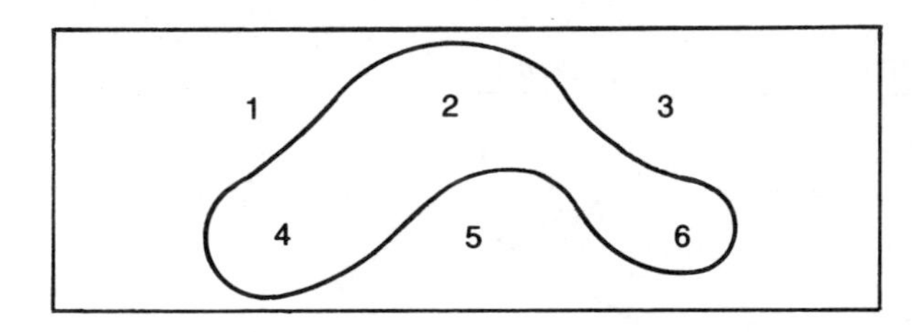

Fig. 5.8 Venn diagram for the six elements when a die is thrown, and the set (2, 4, 6).

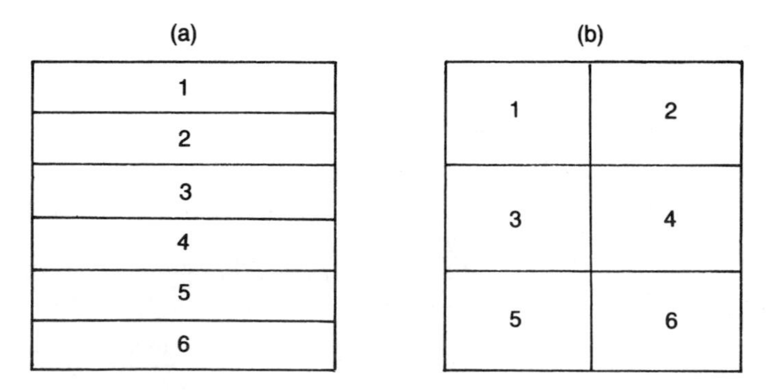

Fig. 5.9 Rees diagrams for the outcomes of the throw of a die.

(a)

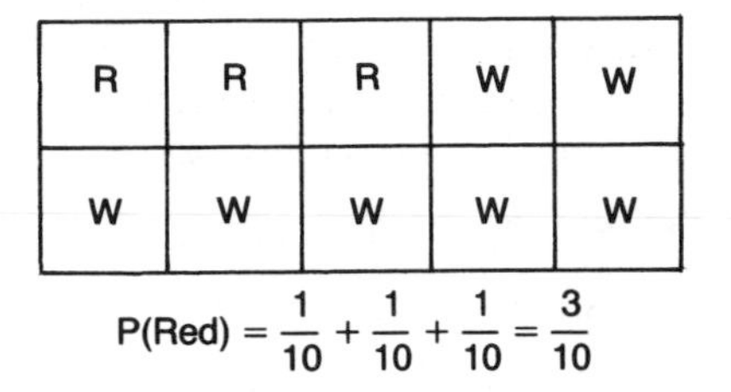

(b)

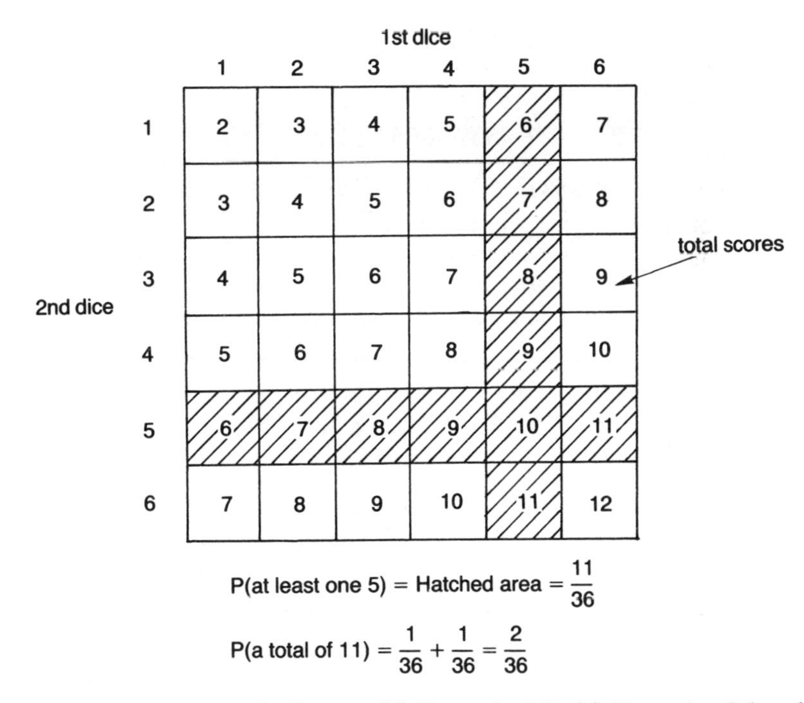

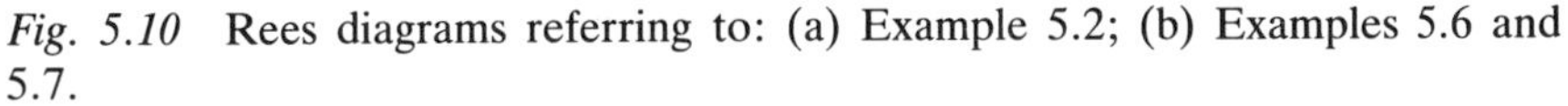

Fig. 5.10 Rees diagrams referring to: (a) Example 5.2; (b) Examples 5.6 and 5.7.

5.15 SUMMARY

Probability as a measure of uncertainty may be defined using the *a priori* and relative frequency definitions. The first is useful in games of chance, the second when we have sufficient experimental data.

In calculating probabilities involving more than one event, two laws of probability are useful. The multiplication law can be expressed as:

$$P(E_1 \text{ and } E_2) = P(E_1) \times P(E_2|E_1)$$

For statistically independent events, this reduces to:

$$P(E_1 \text{ and } E_2) = P(E_1) \times P(E_2)$$

The addition law can be written as:

$$P(E_1 \text{ or } E_2 \text{ or both}) = P(E_1) + P(E_2) - P(E_1 \text{ and } E_2)$$

For mutually exclusive events, this reduces to:

$$P(E_1 \text{ or } E_2 \text{ or both}) = P(E_1) + P(E_2)$$

Various graphical methods can be very helpful in representing the outcomes and their associated probabilities, for small experiments. Of these diagrams, the probability tree is one of the most helpful.

WORKSHEET 5: PROBABILITY

Questions 1–7 are based on sections 5.1–5.6.

1. Distinguish between the *a priori* and relative frequency definitions of probability.
2. If the probability of a success outcome of an experiment is 0.2, what is the probability of failure?
3. When two coins are tossed the result can be two heads, one head and one tail, or two tails, and hence each of these events has probability 1/3. What is wrong with this argument? What is the correct argument?
4. A coin is tossed five times. Each time it comes down heads. Hence the probability of heads is 5/5 = 1. Discuss.
5. Three ordinary dice, one yellow, one blue and one green, are placed in a bag. A trial involves selecting one die at random from the bag and rolling it, the colour and score being noted.
   (a) What does 'at random' mean here?
   (b) Write down the set of all possible outcomes.
   (c) Are the outcomes equally likely?
   (d) What is the probability of each outcome?
   (e) What are the probabilities of the following events:
      (i) Yellow with any score?
      (ii) Yellow with an even score?
      (iii) Even score with any colour?
      (iv) Yellow 1 or blue 2 or green 3?
      (v) Neither even blue nor odd yellow?
6. For the 27 female students whose heights are listed in Appendix A, draw a histogram like Fig. 3.4. If one female student is selected at random, what is the probability that her height will be:
   (a) Between 164.5 and 169.5 cm?

(b) Between 149.5 and 179.5 cm?
Express your answer to (a) as the ratio of two areas of your histogram.

7. If an ordinary drawing pin is tossed in the air it can fall in one of two ways: point upwards, which we will call event U, or point downwards, which we will call event U'. As it is not possible to obtain an *a priori* estimate of the probability that the drawing pin will fall point upwards, $P(U)$, we can estimate this probability by carrying out an experiment as follows: Toss a drawing pin 50 times and record the result of each of 50 trials as U or U'. Use the relative frequency definition to estimate $P(U)$ after 1, 2, 3, 4, 5, 10, 20, 30, 40 and 50 trials. Plot a graph of $P(U)$ versus the number of trials. It should indicate that the estimated probability fluctuates less as more trials are performed.

Questions 8–22 are based on sections 5.8–5.15.

8. Write down the following events in symbol form, where A and B are two events: (a) not A, (b) A given B, (c) B given A.
9. What is meant by: (a) $P(A|B)$, (b) $P(B|A)$, (c) $P(A')$, (d) A and B are statistically independent, (e) A and B are mutually exclusive? For (d) and (e), think of examples.
10. What is the 'and' law of probability, as applied to events A and B? What happens if A and B are statistically independent?
11. What is the 'or' law of probability, as applied to events A and B? What happens if A and B are mutually exhaustive?
12. What can be concluded if, (a) $P(A|B) = P(A)$, (b) $P(A \text{ and } B) = 0$?
13. What is the probability of a 3 or a 6 with one throw of a die?
14. What is the probability of a red card, a picture card (ace, King, Queen or Jack), or both, when a card is drawn from a pack at random?
15. A coin is tossed three times. Before each toss a subject guesses the result as 'head' or 'tails'. If the subject always guesses 'tails', what is the probability that the subject will be correct: (a) three times, (b) twice, (c) once, (d) no times? Hint: draw a probability tree.
16. Three marksmen have probabilities 1/2, 1/3 and 1/4 of hitting a target with each shot. If all three marksmen fire simultaneously, calculate the probability that at least one will hit the target. (Refer to Example 5.9 if necessary.)
17. Of the sparking plugs manufactured by a firm, 3% are defective. In a

random sample of four plugs, what is the probability that exactly one will be defective?

18. Suppose that, of a group of people, 30% own both a house and a car, 40% own a house and 70% own a car. What proportion: (a) own at least a house or a car? (b) of car owners are also house-holders?

19. Of 14 double-bedded rooms in a hotel, 9 have a bathroom. Of 6 single-bedded rooms, 2 have a bathroom.
    (a) What is the probability that, if a room is randomly selected, it will have a bathroom?
    (b) If a room is selected from those with a bathroom, what is the probability that it will be a single room?

20. It is known from past experience of carrying out surveys in a certain area that 25% of the houses will be unoccupied on any given day. In a proposed survey it is planned that houses unoccupied on the first visit will be visited a second time. One member of the survey team calculated that the proportion of houses occupied on either the first or second visit is $1 - (\frac{1}{4})^2 = 15/16$. Another member of the team calculated this proportion as $1 - 2 \times 1/4 = 1/2$.

    Show that both members' arguments are incorrect. Give the correct argument.

21. A two-stage rocket is to be launched on a space mission. The probability that the lift-off will be a failure is 0.1. If the lift-off is successful the probability that the separation of the stages will be a failure is 0.05. If the separation is successful, the probability that the second stage will fail to complete the mission is 0.03.

    What is the probability that the whole mission will: (a) be a success? (b) be a failure?

22. If one student is selected at random from the 40 listed in Appendix A, what is the probability that this student is:
    (a) Male?
    (b) Female?
    (c) At least 165 cm in height?
    (d) At least 165 cm in height, given that the student is (i) male, (ii) female?
    (e) Male, given that the student is (i) at least 165 cm in height, (ii) less than 165 cm in height? Do you think that sex is independent of height?
    (f) Male and studying for a BSc?
    (g) Male, or studying for a BSc, or both male and studying for a BSc?

Discrete probability distributions

6.1 INTRODUCTION

If a discrete variable can take values with associated probabilities it is called a **discrete random variable**. The values and the probabilities are said to form a **discrete probability distribution**. For example, the discrete probability distribution for the variable 'number of heads resulting from three tosses of a fair coin', may be represented as in Table 6.1. These probabilities may be determined by the use of a probability tree (see Worksheet 5, Question 15).

Table 6.1 Probability distribution for the number of heads in three tosses of a coin

Number of heads	0	1	2	3
Probability	0.125	0.375	0.375	0.125

There are several standard types of discrete probability distribution. We will consider two of the most important, namely, the binomial distribution (sections 6.2–6.8) and the Poisson distribution (sections 6.9–6.16).

6.2 BINOMIAL DISTRIBUTION, AN EXAMPLE

Consider another coin-tossing example, but this time we will toss the coin 10 times. The number of heads will vary if we repeatedly toss the coin and we can note the following:

1. We have a fixed number of tosses, namely 10.
2. Each toss can result in one of only two outcomes, namely heads and tails.

3. The probability of heads is the same for each toss, and is 0.5 for a fair coin.
4. The tosses are independent in the sense that the probability of heads for any toss is unaffected by the result of any previous toss.

Because these four conditions are satisfied, the experiment of tossing the coin 10 times is an example of what is called a **binomial experiment** (consisting of 10 so-called **Bernoulli trials**).

The variable 'number of heads in the 10 tosses of a coin' is said to have a **binomial distribution** with **parameters** 10 and 0.5, which we write in short-hand form as $B(10, 0.5)$. The first parameter, 10, is the number of trials or tosses; the second parameter, 0.5, is the probability of heads in each single trial or toss.

6.3 THE GENERAL BINOMIAL DISTRIBUTION

In order to generalize the example of the previous section, the outcomes of each trial in a binomial experiment are conventionally referred to as 'success' (one of the outcomes) and 'failure' (the other outcome). The general binomial distribution is denoted by $B(n, p)$, where the parameters n and p are, respectively, the number of trials and the probability of success in a single trial.

In order to decide *a priori* whether a variable has a binomial distribution, we must check the following four conditions (generalizing on those of the previous section):

1. There must be a fixed number of trials, n.
2. Each trial can result in one of only two outcomes, which we refer to as success and failure.
3. The probability of success in a single trial, p, is constant.
4. The trials are independent, so that the probability of success in any trial is unaffected by the result of any previous trials.

If all four conditions are satisfied, then the discrete random variable, which we call x, to stand for 'the number of successes in n trials', has a $B(n, p)$ distribution.

Unless n is small (at most 3) the methods we used in Chapter 5 are inefficient for calculating probabilities. Luckily we can use a formula, which we shall quote without proof, or in some cases we can use statistical tables (see section 6.6) or Minitab (see section 6.7) to find probabilities for a particular binomial distribution, i.e. if we know the numerical values of n and p.

The formula for the binomial distribution is:

$$P(x) = \binom{n}{x} p^x(1 - p)^{n-x} \qquad \text{for } x = 0, 1, 2, \ldots, n$$

This formula is not difficult to use if each part is understood separately:

P(x) means 'the probability of x successes in n trials'.

$\binom{n}{x}$ is a shorthand for $n!/x!(n - x)!$, where $n!$ means 'factorial n' (refer to section 2.2 if necessary).

$x = 0, 1, 2, \ldots, n$ means that we can use this formula for each of these values of x, which are the possible numbers of successes in n trials.

6.4 CALCULATING BINOMIAL PROBABILITIES, AN EXAMPLE

For the example of tossing a coin 10 times, $n = 10$. If we assume that the coin is fair and we regard 'heads' as the outcome we will think of as a 'success', then $p = 0.5$. We can now calculate the probabilities of getting all possible numbers of heads using the formula:

$$P(x) = \binom{10}{x}(0.5)^x(1 - 0.5)^{10-x} \qquad \text{for } x = 0, 1, 2, \ldots, 10.$$

For example, to find the probability of getting 8 heads in 10 tosses we put $x = 8$ in this formula:

$$\begin{aligned} P(8) &= \binom{10}{8}(0.5)^8(1 - 0.5)^{10-8} \\ &= \frac{10!}{8!2!}(0.5)^8(0.5)^2 \\ &= 0.0439. \end{aligned}$$

Proceeding similarly for all other possible values of x, and recalling that $0! = 1$, we can obtain Table 6.2.

Since this set of 11 possible events resulting from tosses of a coin 10 times form a mutually exclusive and exhaustive set, the probabilities in Table 6.2 sum to 1 (recalling section 5.11). The probabilities for any discrete distribution should sum to 1 in this way (apart from rounding errors, which may affect the third decimal place), and this fact provides a useful check on our calculations.

The information in Table 6.2 can also be presented graphically (see Fig. 6.1) in a graph which is similar to the line chart in Fig. 3.7, except that the vertical axis represents probability rather than frequency.

Notes

(a) The distribution is symmetrical about the centre line at $x = 5$, but this is because $p = 0.5$ for this distribution, which is exactly in the middle

Table 6.2 Probabilities for a binomial distibution for $n = 10$, $p = 0.5$

Number of heads, x	0	1	2	3	4	5	6	7	8	9	10
Probabilty, P(x)	0.001	0.010	0.044	0.117	0.205	0.246	0.205	0.117	0.044	0.010	0.001

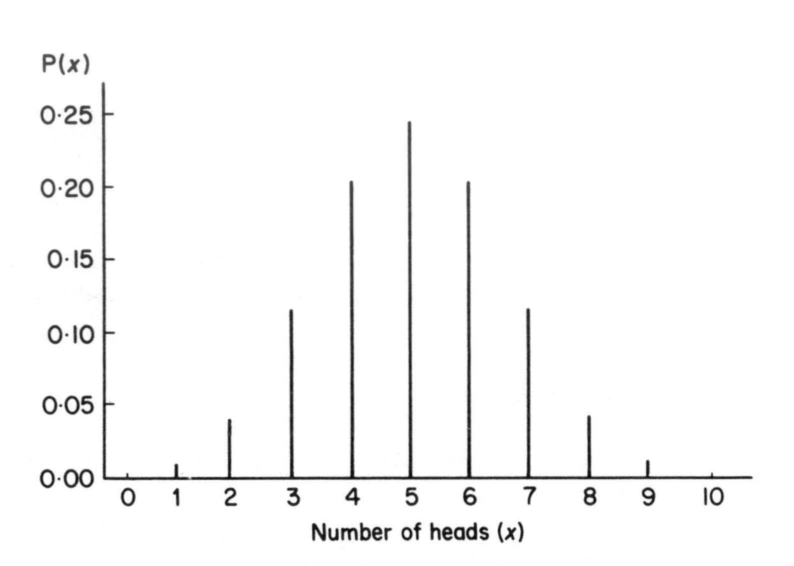

Fig. 6.1 Probabilities for a binomial distribution with $n = 10$, $p = 0.5$.

of the possible range of 0 to 1 for a probability. Binomial distributions with p values greater than 0.5 will be negatively skew, while those with p values less than 0.5 will be positively skew.

(b) Since the number of heads is a discrete random variable, Fig. 6.1 is also discrete with gaps between the possible values.

6.5 THE MEAN AND STANDARD DEVIATION OF THE BINOMIAL DISTRIBUTION

In Chapter 4 we calculated the mean and standard deviation of sample data. In a similar way we can think of the mean and standard deviation of a binomial distribution, meaning the mean and standard deviation of the value taken by the variable in repeated binomial experiments. Instead of having to carry out these experiments in order to calculate the mean and standard deviation, it can be shown mathematically that the following formulae may be applied to any binomial distribution:

$$\text{mean} = np, \quad \text{standard deviation} = \sqrt{(np(1-p))}$$

Example

For the coin-tossing example of the previous section, $n = 10$, $p = 0.5$, so that the mean equals $10 \times 0.5 = 5$, and the standard deviation equals $\sqrt{(10 \times 0.5 \times 0.5)} = 1.58$.

It should seem intuitively reasonable that the mean number of heads in repetitions of 10 tosses of a coin is 5. But what does a standard deviation of 1.58 tell us? As stated before, we shall derive more meaning from the value of a standard deviation when we discuss the normal distribution in Chapter 7. For the time being we remember that the larger the standard deviation, the more the variable will vary. So a variable with a $B(20, 0.5)$ distribution varies more than a variable with a $B(10, 0.5)$ distribution, since their respective standard deviations are 2.24 and 1.58.

6.6 BINOMIAL PROBABILITIES USING TABLES

In order to save time in calculating binomial probabilities, Table D.1 of Appendix D may be used, instead of the formula of P(x), for certain values of n and p. Table D.1 gives cumulative probabilities, that is the probabilities of 'so many or fewer successes'.

Example

$n = 10$, $p = 0.5$. In Table D.1, find the column of probabilities for these values of n and p.

To find P(8) = P(8 successes in 10 trials, $p = 0.5$), find the row labelled $r = 8$, and read that:

$$\text{P(8 or fewer successes in 10 trials, } p = 0.5) = 0.9893$$

Now find the row labelled $r = 7$ and read that:

$$\text{P(7 or fewer successes in 10 trials, } p = 0.5) = 0.9453$$

Subtracting,

$$\text{P(exactly 8 successes in 10 trials, } p = 0.5) = 0.0440.$$

This agrees to 3 dps with the answer obtained in section 6.4 using the formula.

In general, we use the idea that:

$$\text{P}(x \text{ successes}) = \text{P}(x \text{ or fewer successes}) - \text{P}((x-1) \text{ or fewer successes}).$$

Example

$n = 10, p = 0.5$. Find P(10).

$$P(10) = P(10 \text{ or fewer}) - P(9 \text{ or fewer}) = 1 - 0.9990 = 0.0010.$$

Because only certain values of n and p are given in the tables (since the tables would need to be very extensive to cover all possible values of n and p) you should learn both methods unless you have access to Minitab (see section 6.7). The fact that no p-values above 0.5 are given in Table D.1 is less of a restriction than it might at first appear to be. This is because we can always treat as our 'success' the outcome which has the smaller probability. The following example illustrates this point and some other logical points.

Example

If 70% of male students eventually marry, what is the probability that 40 or more of a random sample of 50 male students will eventually marry? The outcome with the smaller probability is 'not marry', so we treat 'not marry' as 'success' and 'marry' as 'failure'. Assuming that the four binomial conditions apply, use Table D.1 for $n = 50$, $p = 1 - 0.7 = 0.3$, $r = 10$.

$$\begin{aligned} P(40 \text{ or more out of 50 will marry}) &= P(40 \text{ or more 'failures'}) \\ &= P(10 \text{ or fewer 'successes'}) \\ &= 0.0789 \end{aligned}$$

By the way, if you think this is a tortuous calculation, think of the alternative. Using the formula for P(x), this time take 'marry' as 'success' and use $n = 50$, $p = 0.7$. Thus:

$$\begin{aligned} P(40 \text{ or more out of 50 will marry}) &= P(40) + P(41) + \cdots + P(50) \\ &= \binom{50}{40} 0.7^{40}(1 - 0.7)^{10} \\ &\quad + \binom{50}{41} 0.7^{41}(1 - 0.7)^{9} + \cdots \end{aligned}$$

which is a rather tedious calculation!

6.7 BINOMIAL PROBABILITIES USING MINITAB

Minitab can be used to generate probabilities for the binomial distribution. The PDF command produces probabilities of exact numbers of successes for a specified binomial distribution, while the CDF command produces cumulative probabilities of so many or fewer successes. Examples are given in Tables 6.3 and 6.4. Notice that the probabilities in Table 6.3 are equal to those in Table 6.2, apart from rounding. Also note

Table 6.3 Probabilities for the $B(10, 5)$ distribution

MTB> PDF;
SUBC> BINOMIAL N = 10, P = 0.5.
BINOMIAL WITH N = 10 P = 0.5

K	P (X = K)
0	0.0010
1	0.0098
2	0.0439
3	0.1172
4	0.2051
5	0.2461
6	0.2051
7	0.1172
8	0.0439
9	0.0098
10	0.0010

Table 6.4 Cumlative probabilities for the $B(10, 0.5)$ distribution

MTB> CDF;
SUBC> BINOMIAL N = 10, P = 0.5.
BINOMIAL WITH N = 10 P = 0.5

K	P (X LESS or = K)
0	0.0010
1	0.0107
2	0.0547
3	0.1719
4	0.3770
5	0.6230
6	0.8281
7	0.9453
8	0.9893
9	0.9990
10	1.0000

that the probabilities in Table 6.4 are identical to those in Table D.1 for $n = 10$, $p = 0.5$.

6.8 SIMULATION OF BINOMIAL DISTRIBUTIONS USING MINITAB

It is also possible to use Minitab to simulate binomial distributions. For example, the $B(10, 0.5)$ distribution is a model for the binomial experi-

Table 6.5 100 simulations of a $B(10, 0.5)$ experiment

```
MTB> RANDOM 100  C1;
SUBC> BINOMIAL  N = 10, P = 0.5.
MTB> PRINT  C1
C1
3  6  6  6  8  6  9  6  4  6  6  6  6  6  6
6  6  5  6  5  8  3  6  7  3  3  7  5  6  6
5  7  6  7  3  4  6  4  6  6  3  6  8  5  4
6  5  5  8  4  4  5  6  3  7  2  5  3  6  6
5  3  3  8  5  2  4  5  5  5  2  4  7  6  7
5  3  5  3  5  5  7  5  5  6  6  6  5  3  5
4  5  4  7  5  5  4  6  8  6
```

ment of tossing a coin 10 times and counting the number of heads. This experiment may be simulated 100 times, say, by the commands shown in Table 6.5, which also shows the results of the simulation.

If this experiment is carried out a large number of times, theory (see section 6.5) suggests that the mean and standard deviation of the number of successes should be 5 and 1.58, respectively for this example. Even with only 100 repetitions the DESCRIBE C1 command (not shown in Table 6.5) gives a mean of 5.26 and standard deviation of 1.495, quite close to their 'theoretical' values. We could also compare the shape of the above data (using HISTOGRAM C1) with the 'theoretical' shape shown in Fig. 6.1.

6.9 POISSON DISTRIBUTION, AN INTRODUCTION

The second standard type of discrete probability distribution we will consider, the **Poisson distribution**, is concerned with the variable 'number of random events per unit time or space'. The word 'random' in this context implies that there is a constant probability that the event will occur in one unit of time or space (space can be one-, two- or three-dimensional).

6.10 SOME EXAMPLES OF POISSON VARIABLES

There are many examples which may be used to illustrate the great variety of applications of the Poisson distribution as a 'model' for random events:

(a) At the telephone switchboard in a large office block there may be a constant probability that a telephone call will be received in a given minute. The number of calls received per minute will therefore have a Poisson distribution.

(b) In spinning wool into a ball from the raw state, there may be a constant probability that the spinner will have to stop to untangle a knot. The number of stops per 100 metres of finished wool will then have a Poisson distribution.

(c) In the production of polythene sheeting there may be a constant probability of a blemish (called 'fish-eyes') which makes the film unsightly or opaque. The number of blemishes per square metre will then have a Poisson distribution.

Other examples concerned with random events in time are the number of postilions killed by lightning in the days of horse-drawn carriages, the number of major earthquakes recorded per year, the number of α-particles emitted per unit time by a radioactive source and the number of cases of childhood leukaemia per year per unit area (both close to and far from nuclear sites).

6.11 THE GENERAL POISSON DISTRIBUTION

In order to decide whether a discrete random variable has a Poisson distribution we must be able to answer 'yes' to the following questions:

1. Are we interested in random events per unit time or space?
2. Is the number of events which might occur in a given unit time or space theoretically unlimited?

If the answer to the first question is yes, but the answer to the second question is no, the distribution may be binomial – check the four conditions in section 6.3.

In order to calculate the probabilities for a Poisson distribution we can use a formula, which we shall quote without proof, or, in some cases, we can use tables (see section 6.14) or we can use Minitab (see section 6.15) if we know the numerical value of the parameter m (defined below) for a particular Poisson distribution. The formula is

$$P(x) = \frac{e^{-m}m^x}{x!} \qquad \text{for } x = 0, 1, 2, \ldots$$

Here $P(x)$ means the probability of x random events per unit time or space, e is the number 2.718 . . . (refer to section 2.4 if necessary), m is the mean number of events per unit time or space, and $x = 0, 1, 2, \ldots$ means that we can use this formula for $x = 0$ or any positive whole number.

6.12 CALCULATING POISSON PROBABILITIES, AN EXAMPLE

Telephone calls arrive at a switchboard at the rate of 1 per minute on average. What are the probabilities that 0, 1, 2, calls will be received in a period of 5 minutes?

Since the probabilities of interest relate to a unit of time of 5 minutes we calculate m as the mean number of calls per 5 minutes. So $m = 5$ for this example, and

$$P(x) = \frac{e^{-5}5^x}{x!} \quad \text{for } x = 0, 1, 2, \ldots$$

For example, for $x = 6$ calls in 5 minutes,

$$P(6) = \frac{e^{-5}5^6}{6!} = 0.146$$

Substituting other values of x, and recalling that $0! = 1$, we obtain Table 6.6.

These 11 possible events (in observing a switchboard for 5 minutes) form a mutually exclusive set but they are not an exhaustive set since x-values above 10 are possible, although their probabilities are small. The probabilities in Table 6.6 sum to 0.986, which illustrates the same point.

The information in Table 6.6 is also presented graphically in Fig. 6.2 (in a graph similar to Fig. 6.1 for a binomial distribution). Note that the distribution is slightly positively skewed. Poisson distributions with values of m less than 5 will be more positively skewed, while those with values of m greater than 5 will be less skewed and hence more symmetrical.

6.13 THE MEAN AND STANDARD DEVIATION OF THE POISSON DISTRIBUTION

The mean of the Poisson distribution is m, as already stated in section 6.9, and the standard deviation of the Poisson distribution is $\sqrt{m}$.

For the example above, the mean is 5 and the standard deviation is 2.24.

Table 6.6 Probabilities for a Poisson distribution for $m = 0.5$

Number of calls received in 5 minutes (x)	0	1	2	3	4	5	6	7	8	9	10
Probability P(x)	0.007	0.034	0.084	0.140	0.176	0.176	0.146	0.104	0.065	0.036	0.018

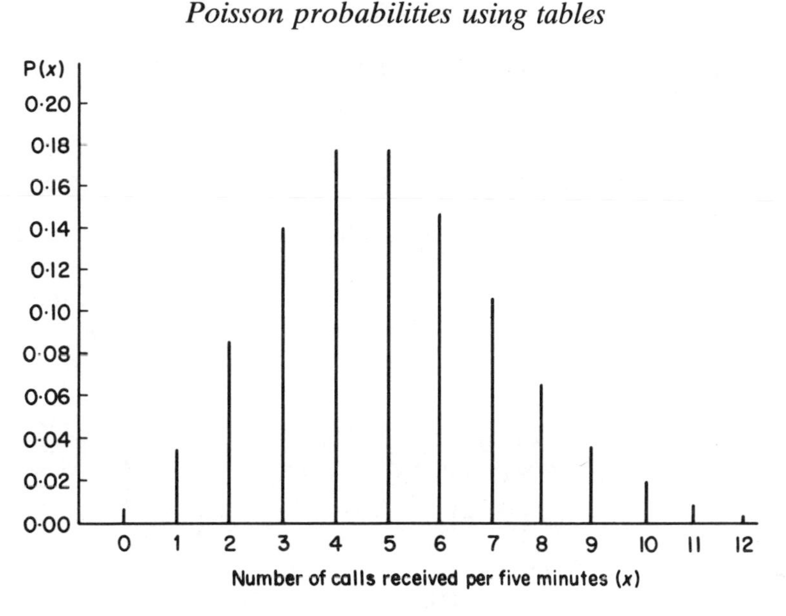

Fig. 6.2 Probabilities for a Poisson distribution with $m = 5$.

6.14 POISSON PROBABILITIES USING TABLES

To save time in calculating Poisson probabilities, Table D.2 of Appendix D may be used for certain values of m, instead of the formula for $P(x)$. In a similar way to Table D.1 of Appendix D, Table D.2 gives cumulative probabilities, that is the probabilities of 'so many or fewer random events per unit time or space'.

Example (the same as that in section 6.12)

$m = 5$. In Table D.2 find the column of probabilities for this value of m.

To find $P(6) = P(6 \text{ random events when } m = 5)$, find the row labelled $r = 6$, and read that:

$$P(6 \text{ or fewer random events when } m = 5) = 0.7622$$

Now find the row labelled $r = 5$ and read that:

$$P(5 \text{ or fewer random events when } m = 5) = 0.6160$$

Subtracting,

$$P(6 \text{ random events when } m = 5) = 0.1462$$

This agrees to 3 dps with the answer obtained in section 6.12 using the formula.

In general we use the idea that:

$$\text{P}(x \text{ random events}) = \text{P}(x \text{ or fewer random events}) - \text{P}((x-1) \text{ or fewer random events})$$

6.15 POISSON PROBABILITIES USING MINITAB

Minitab can be used to generate probabilities for the Poisson distribution. The PDF command produces probabilities of exact numbers of random events in time or space for a specified Poisson distribution, while the CDF command produces cumulative probabilities of so many or fewer random events.

Try

```
MTB> PDF;
SUBC> POISSON  MEAN = 5.
```

and compare with Table 6.6. Also try

```
MTB> CDF;
SUBC> POISSON  MEAN = 5.
```

and compare with Table D.2 for $m = 5$.

6.16 SIMULATION OF POISSON DISTRIBUTIONS USING MINITAB

It is also possible to simulate Poisson distributions using Minitab. For example, if we know that the mean number of randomly arriving telephone calls at a switchboard is 5 calls per 5 minutes, we may simulate 100 periods each of 5 minutes' duration and print the results with the commands:

```
MTB> RANDOM 100 C1;
SUBC> POISSON  MEAN = 5.
MTB> PRINT C1
```

The theoretical values for the mean and standard deviation of the number of calls are $m = 5$ and $\sqrt{m} = 2.24$ (see section 6.13). The agreement between the values obtained from the simulation and the theoretical values should improve as the number of repetitions increases.

6.17 POISSON APPROXIMATION TO THE BINOMIAL DISTRIBUTION

There are examples of binomial distributions for which the calculation of approximate probabilities is made easier by the use of the formula or tables for the Poisson distribution! Such an approach can be justified theoretically for binomial distributions with large values of n and small values of p. The resulting probabilities are only approximate, but quite good approximations may be obtained when $p < 0.1$, even if n is not large, by putting $m = np$.

Example

Assume that 1% of people are colour-blind. What is the probability that 10 or more of a random sample of 500 people will be colour-blind?

This is a binomial problem with $n = 500$ and $p = 0.01$. But Table D.1 cannot be used for $n = 500$, and to use the binomial formula we would need to calculate $1 - P(0) - P(1) - \cdots - P(9)$.

However, using the Poisson approximation for $m = np = 5$, and Table D.2 for $r = 9$, we read that:

$$P(\text{9 or fewer colour-blind in a sample of 500 people}) = 0.9682$$

Hence

$$\begin{aligned} P(\text{10 or more colour-blind in a sample of 500 people}) &= 1 - 0.9682 \\ &= 0.0318. \end{aligned}$$

An alternative approach to this example would simply be to use Minitab to generate binomial probabilities as in section 6.7. However, it is, I feel, important to see connections between distributions if they can be demonstrated reasonably easily. Otherwise there is a danger that each distribution is seen as a different 'rabbit' pulled out of a hat.

6.18 SUMMARY

The binomial and the Poisson distributions are two of the most important discrete probability distributions. The binomial distribution gives the probabilities for the numbers of successes in a number of trials, if four conditions hold. Binomial probabilities may be obtained using:

$$P(x) = \binom{n}{x} p^x (1 - p)^{n-x}$$

or, in certain cases, Table D.1, or Minitab.

The Poisson distribution gives the probabilities for the number of random events per unit time or space. Poisson probabilities may be calculated using:

$$P(x) = \frac{e^{-m} m^x}{x!}$$

or, in certain cases, Table D.2, or Minitab.

If $p < 0.1$, it may be preferable to calculate binomial probabilities using the Poisson approximation.

WORKSHEET 6: THE BINOMIAL AND POISSON DISTRIBUTIONS

Questions 1–12 are on the binomial distribution.

1. For the binomial distribution, what do n and p stand for?
2. How can you tell *a priori* whether a discrete random variable has a binomial distribution?
3. In a binomial experiment each trial can result in one of two outcomes. Which one shall I call a 'success' and which a 'failure'?
4. What is the general name for the variable which has a binomial distribution?
5. For the distribution $B(3, 0.5)$,
   (a) How many outcomes are there to each trial?
   (b) How many trials are there?
   (c) How many possible values can the variable take?
   (d) What is the mean and what is the standard deviation for this distribution?
   (e) Is this distribution symmetrical?
6. For families with four children what are the probabilities that a randomly selected family will have 0, 1, 2, 3 or 4 boys, assuming that boys and girls are equally likely at each birth? Check that the probabilities sum to 1. Why do they?

   Given 200 families each with four children, how many would you expect to have 0, 1, 2, 3, or 4 boys?
7. In a multiple-choice test there are five possible answers to each of 20 questions. If a candidate guesses the answer to each question:
   (a) What is the mean number of correct answers you would expect the candidate to obtain?
   (b) What is the probability that the candidate will pass the test by getting 8 or more correct answers?
   (c) What is the probability that the candidate will get at least one correct answer?
8. In a large batch 5% of items are defective. If 50 items are selected at random, what is the probability that:
   (a) At least one will be defective?
   (b) Exactly two will be defective?
   (c) Ten or more will be defective?

Use tables to answer these questions initially, but check the answers to (a) and (b) using a formula.

9. In an experiment with rats, each rat goes into a T-maze in which there is a series of T-junctions. At each junction a rat can either turn left or right. Assuming a rat chooses at random, what are the probabilities that it will make 0, 1, 2, 3, 4 or 5 right turns out of 5 junctions?

10. A new method of treating a disease has a 70% chance of resulting in a cure. Show that if a random sample of 10 patients suffering fom the disease are treated by this method the chance that there will be more than 7 cures is about 0.38. Check this answer. What other word could be used in place of 'chance'?

11. Fifty grams of yellow wallflower seeds are thoroughly mixed with 200 g of red wallflower seeds. The seeds are then bedded out in rows of 20.
    (a) Assuming 100% germination, why should the number of yellow wallflower plants per row have a binomial distribution?
    (b) What are the values of n and p for this distribution?
    (c) What is the probability of getting a row with:
        (i) No yellow wallflower plants in it?
        (ii) One or more yellow wallflower plants in it?

12. A supermarket stocks eggs in boxes of six, and 10% of the eggs are cracked. Assuming that the cracked eggs are distributed at random, what is the probability that a customer will find that the first box he chooses contains:
    (a) No cracked eggs?
    (b) At least one cracked egg?
    If he examines five boxes, what is the probability that three or more will contain no cracked eggs?

Questions 13–19 are on the Poisson distribution.

13. For the Poisson distribution we use the formula

$$P(x) = \frac{e^{-m}m^x}{x!}$$

    What do the symbols m, e, x stand for? What values can x take?

14. If a variable has a Poisson distribution with a mean of 4, what is its standard deviation and what is its variance? What can you say about the mean and variance of any Poisson distribution?

15. The Poisson distribution is the distribution of the number of random events per unit time. What does the word 'random' mean here?

16. Assuming that breakdowns in a certain electricity supply occur randomly with a mean of one breakdown every 10 weeks, calculate the probabilities of 0, 1, 2 breakdowns in any period of one week.

17. Assume that the number of misprints per page of a book has a Poisson distribution with a mean of one misprint per five pages. What percentage of pages contain no misprints? How many pages would you expect to have no misprints in a 500-page book?

18. A hire firm has three ladders which it hires out by the day. Records show that the mean demand over a period is 2.5 ladders per day. If it is assumed that the demand for ladders follows a Poisson distribution, find:
    (a) The pecentage of days on which no ladder is hired.
    (b) The percentage of days on which all three ladders are hired.
    (c) The percentage of days on which demand outstrips supply.

19. A roll of cloth contains an average of 3 defects per 100 square metres distributed at random. What is the probability that a randomly chosen section of 100 square metres of cloth contains:
    (a) No defects?
    (b) Exactly three defects?
    (c) Three or more defects?

Questions 20–22 are on the Poisson approximation to the binomial distribution.

20. A rare blood group occurs in only 1% of the population, distributed at random. What is the probability that at least one person in a random sample of 100 has blood of this group? Use both the 'binomial' method and the 'Poisson approximation to the binomial' method. Compare your answers. Which is correct?

21. If, in a given country, an average of one miner in 2000 loses his life due to accident per year, calculate the probability that a mine in which there are 8000 miners will be free from accidents in a given year.

22. The average number of defectives in batches of 50 items is 5. Find the probability that a batch will contain:
    (a) 10 or more defectives.
    (b) Exactly five defectives.

    Use both the binomial and the Poisson approximation to the binomial methods, and compare your answers.

Continuous probability distributions

7.1 INTRODUCTION

In Chapter 3 we considered an example of a continuous variable, namely the heights of students, and we summarized the heights of 27 female students by a histogram (see Worksheet 5, Question 6), reproduced here as Fig. 7.1. We also saw in that question how to express a probability as the ratio of two areas, so that we could make statements such as

$$\text{P(randomly selected female student has a height between 164.5 and 169.5 cm)} = \frac{\text{Area of rectangle on base 164.5–169.5}}{\text{Total area of histogram}}$$

Suppose we apply this idea to the heights of all female students in higher education. In the histogram the number of students in each group would be much greater so we could have many more than six groups, and still have a fairly large number of students in each group. This histogram would look something like Fig. 7.2, where the vertical sides of the rectangles have been omitted and the tops of the rectangles have been smoothed into a curve.

If this graph is 'scaled' in the vertical direction so that the total area under the curve is 1 in some units, then we would be wrong to keep calling the vertical axis 'Number of students'. However, this curve would have the property that the probability of a female student's height being between any two values would be equal to the area under the curve between these values (Fig. 7.3).

For example,

$$\text{P(randomly selected female student has a height between 164.5 and 169.5 cm)} = \text{Area under curve between 164.5 and 169.5}$$

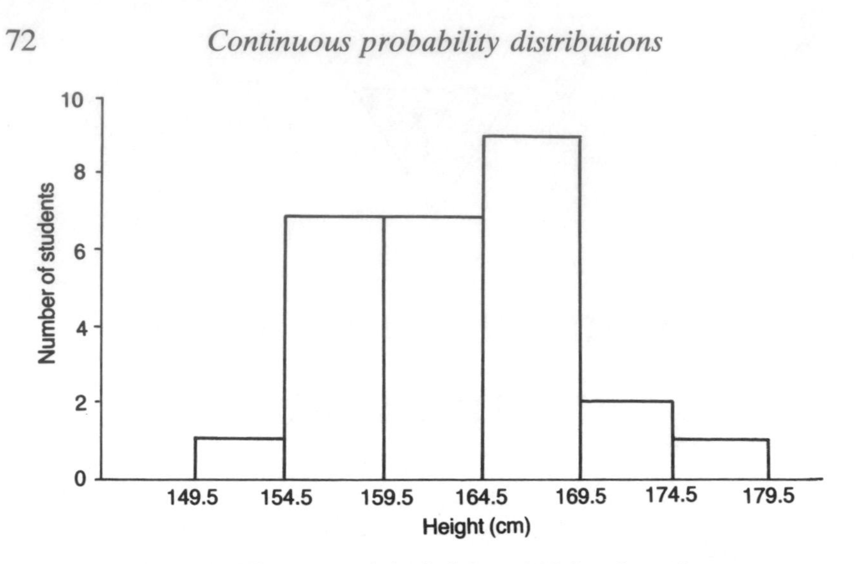

Fig. 7.1 Histogram of the heights of 27 female students.

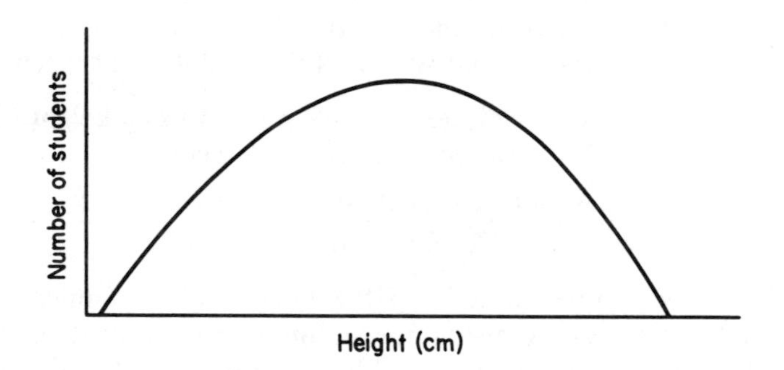

Fig. 7.2 Histogram of the heights of all female students in higher education.

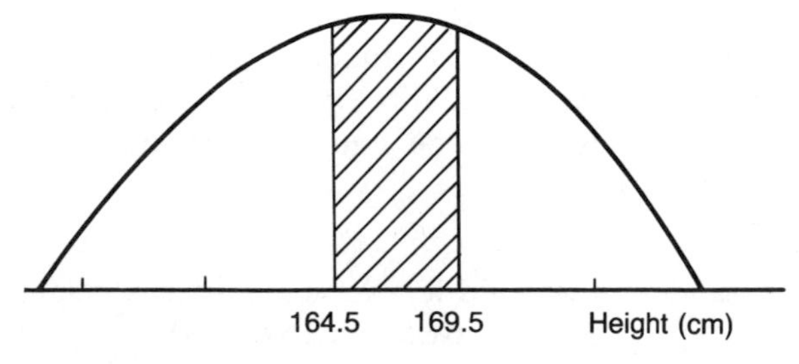

Fig. 7.3 Continuous probability distrubution for the variable 'height'.

Assuming such a curve could be drawn, this is an example of the graphical representation of a **continuous probability distribution**. Compare Fig. 6.1, an example of the graphical representation of a discrete probability distribution.

There are several standard types of continuous probability distribution. We will consider two of the most important, namely the normal distribution and the rectangular (or uniform) distribution.

7.2 THE NORMAL DISTRIBUTION

The normal distribution is the most important in statistics. There are two main reasons for this:

1. It arises when a variable is measured for a large number of nominally identical objects, and when the variation may be assumed to be caused by a number of factors each exerting a small positive or negative random influence on an individual object. An example is the variable 'height of female students', where the variation in heights is caused by many factors such as age, diet, exercise, heights of parents, bone structure and so on.
2. The properties of the normal distribution have a very important application in the statistical theory of drawing conclusions from sample data about the populations from which the samples are drawn (these methods will be discussed from Chapter 8 onwards).

Returning to the idea of graphically representing distributions, the normal distribution has a bell shape (i.e. with most values concentrated towards the centre and few extreme values), is unimodal, and is symmetrical (Fig. 7.4). It has two parameters, μ and σ.

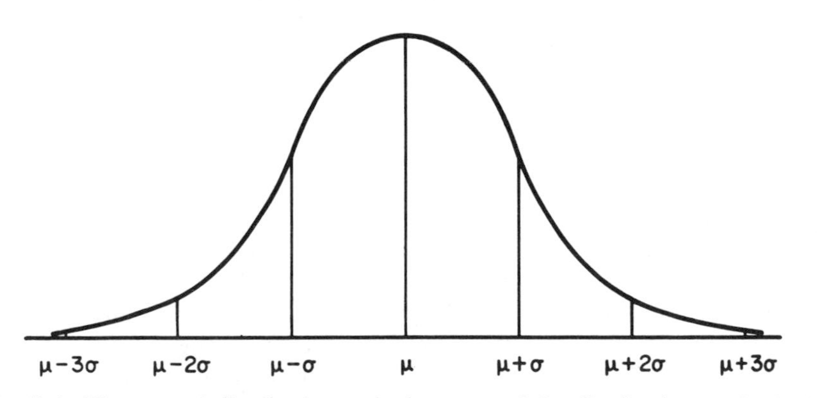

Fig. 7.4 The normal distribution: μ is the mean of the distribution, and σ is the standard deviation of the distribution.

At this point you should note the following important observation concerning notation. In Chapter 4 the symbols $\bar{x}$ and s were used to denote the **sample** mean and **sample** standard deviation, respectively. The Greek symbols μ (lower-case mu) and σ (lower-case sigma) are used here because we are now dealing with a **population** of measurements. Samples and populations are defined and discussed more fully in Chapter 8.

There are a number of related properties of the normal distribution which (at last!) give us a better understanding of the meaning of standard deviation as a measure of variation:

1. Approximately 68% of the area of any normal distribution lies within one standard deviation of the mean. So the area between the vertical lines drawn at $\mu - \sigma$ and $\mu + \sigma$ in Fig. 7.4 is roughly two-thirds of the total area.
2. Approximately 95% of the area of any normal distribution lies within two standard deviations of the mean (exactly 95% of the area lies within 1.96 standard deviations of the mean).
3. Approximately 99.7% of the area of any normal distribution lies within three standard deviations of the mean.

7.3 AN EXAMPLE OF A NORMAL DISTRIBUTION

Suppose that we know that the variable 'height' (of all female students in higher education) is normally distributed with a mean $\mu = 163$ cm and standard deviation $\sigma = 6$ cm.

Using the properties stated in the previous section we could state, for example, that approximately 95% have heights between $163 - 2 \times 6 = 151$ cm and $163 + 2 \times 6 = 175$ cm. This is equivalent to the statement that 'the probability that a randomly selected female student will have a height between 151 and 175 cm is 0.95'.

But how can we calculate probabilities and percentages for other heights of interest? The answer is that we need to be able to obtain areas under any normal distribution curve. We do this using Table D.3(a) of Appendix D; which enables us to calculate probabilities for any normal distribution if we know the numerical values of μ and σ for that distribution. Table D.3(a) actually gives probabilities in terms of areas of the normal distribution curve; that is areas to the left of particular values.

Let us consider the example of the normal distribution of heights given above, i.e. with a mean of $\mu = 163$ cm and standard deviation $\sigma = 6$ cm. (In shorthand form this would be referred to as the $N(163, 6^2)$ distribution, where the general normal distribution is $N(\mu, \sigma^2)$.) Note that all the questions in this section refer to this example.

Question 7.1

What is the probability that a randomly selected female student in higher education will have a height greater than 170 cm?

The answer is the area to the right of 170 in Fig. 7.5, since 'to the right of 170' implies 'greater than 170'.

In order to use Table D.3(a) we first have to 'transform' our normal distribution into one with a mean $\mu = 0$ and standard deviation $\sigma = 1$ (the so-called **standardized** normal distribution). We do this by calculating 'z values' using the formula

$$z = \frac{(x - \mu)}{\sigma}$$

Let us see how to apply this to our example.

Since we are interested in the value 170 cm, let $x = 170$. Now we calculate the z value, using $\mu = 163$ and $\sigma = 6$. Hence:

$$z = \frac{(170 - 163)}{6} = 1.17$$

Using Table D.3(a) for $z = 1.17$ we read that the area to the left of 170 is 0.8790. Hence the area to the right of 170 is $1 - 0.8790 = 0.1210$. We can also state that 12.1% of female students (about 1 in 8) have a height greater than 170 cm (by multiplying by 100).

Question 7.2

What is the probability that height lies between 165 and 170 cm?

The answer is the area between 165 and 170 in Fig. 7.5, which we can think of as:

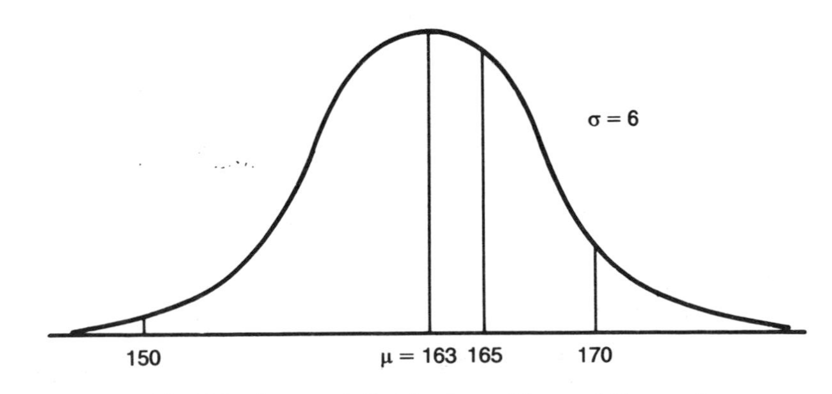

Fig. 7.5 A normal distribution with $\mu = 163$, $\sigma = 6$.

Area to the left of 170 − Area to the left of 165

Since we are now interested in the value 165, let $x = 165$. Now calculate the z value:

$$z = \frac{(165 - 163)}{6} = 0.33$$

Using Table D.3(a) for $z = 0.33$, we read that the area to the left of 165 is 0.6293. Therefore the area between 165 and 170 = 0.8790 − 0.6293 = 0.2497, using the answer to Question 7.1.

The probability that height lies between 165 and 170 is 0.2497. We can also state that 24.97% of female students (about 1 in 4) have a height between 165 and 170 cm.

Question 7.3

What is the probability that height is less than 150 cm?

The answer is the area to the left of 150 in Fig. 7.5. Let $x = 150$, then

$$z = \frac{(150 - 163)}{6} = -2.17$$

The negative sign for z implies what we already know from Fig. 7.5, i.e. that the value 150 lies below the mean of 163. The area given in Table D.3(a) for $z = 2.17$ is 0.9850. Hence the area to the right of $z = 2.17$ is $1 - 0.9850 = 0.015$. By symmetry, this is also the area to the left of $z = -2.17$. So the required probability is 0.015 (1.5% or 1 in about 70).

Question 7.4

What is the probability that height lies between 150 and 165 cm?

From previous answers, the required probability is:

$$0.6293 - 0.0150 = 0.6143$$

Question 7.5

What is the probability that height is less than 163 cm?

By the symmetry of the normal distribution shown in Fig. 7.5, the answer is 0.5, or we could use $x = 163$ and $z = (163 - 163)/6 = 0.$ and Table D.3(a).

7.4 NORMAL PROBABILITIES USING MINITAB

The normal distribution used as a basis for Questions 7.1 to 7.5 was $N(163, 6^2)$. The Minitab command CDF can be used to obtain the probability that a random variable, assumed to follow a given normal distribution, will be less than a specified value. For example, if we wish to find the probability that the height of a female student is less than 170 cm, given a height distribution of $N(163, 6^2)$, the required commands are:

```
MTB> CDF 170;
SUBC> NORMAL  MU = 163, SIGMA = 6.
```

This will produce the response:

```
170    0.8790
```

which implies that the area to the left of 170 cm for the $N(163, 6^2)$ distribution is 0.8790, and this is also the required probability (see previous answer using tables).

The INVCDF command works in reverse. If we know the probability that a normally distributed variable is less than a certain value, we can obtain that value. For example, the commands:

```
MTB> INVCDF  0.8790;
SUBC> NORMAL  MU = 163, SIGMA = 6.
```

will produce the response

```
0.8790    170
```

This gives the same information as in the previous example using the CDF command.

It is a good idea to relate these Minitab outputs to the relevant normal distribution graphically (Fig. 7.5).

We have shown how to tackle Question 7.1 above. Similarly:

(a) Question 7.2 can be answered using MTB> CDF 165; with the same subcommand as above and the answer to Question 7.1.
(b) Question 7.3 can be answered using MTB> CDF 150; with the same subcommand as above.
(c) Question 7.4 can be answered using the answers to previous questions.
(d) Question 7.5 can be answered using MTB> CDF 163; with the same subcommand as above.

7.5 SIMULATION OF THE NORMAL DISTRIBUTION USING MINITAB

It is also possible to make Minitab simulate values from a specified normal distribution. The commands in Table 7.1 will generate 100 random observations from the $N(163, 6^2)$ distribution and put them in C1. These observations should have a mean close to 163 and a standard deviation close to 6 (the more observations the closer you should get to 163 and 6,

Table 7.1 Simulation of 100 random observations from a $N(163, 6^2)$ distribution

```
MTB> RANDOM  100  C1;
SUBC> NORMAL  MU = 163, SIGMA = 6.
MTB> PRINT  C1
MTB> DESCRIBE  C1
```

respectively). You can see how close you get using the DESCRIBE command.

7.6 RECTANGULAR DISTRIBUTION

Because of the great importance of the normal distribution, students who take introductory courses in statistics often believe that all continuous variables are normally distributed. Partly to counteract this erroneous belief at an early stage, we now introduce another continuous probability distribution, namely the **rectangular** (also called the **continuous uniform**) distribution. (The erroneous belief will also be counteracted in Chapter 11 when we deal with inferential methods which deal specifically with non-normal continuous variables.)

The rectangular is a rather dull and flat distribution (Fig. 7.6), but it does have the advantage that probabilities are easy to calculate.

Example

Suppose we consider the 'error' which is made when a person states his or her 'age at last birthday'. The error is the difference

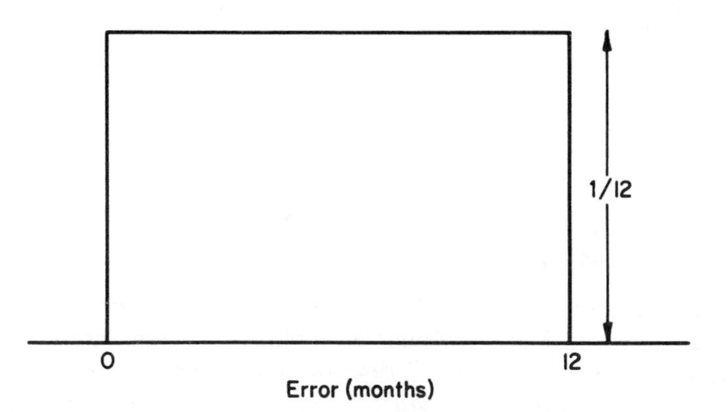

Fig. 7.6 A rectangular distribution for the 'error' in a stated age.

actual age − age at last birthday

and this continuous variable is equally likely to lie anywhere in the range 0 to 12 months, so its probability distribution is as in Fig. 7.6. Note that, since the total area of the rectangle must be equal to 1, the height of the rectangle must be equal to 1/(base) = 1/12.

Question 7.6

What percentage of errors will be less than 3 months?

The probability of an error of less than three months is the area to the left of 3 which is $3 \times 1/12 = 1/4$, so 25% of errors will be less than 3 months.

7.7 THE NORMAL APPROXIMATION TO THE BINOMIAL DISTRIBUTION

Just as there are conditions (section 6.17) when the calculation of approximate binomial probabilities is made easier by using the formula or tables for the Poisson distribution, so there are other conditions when it is preferable to use normal distribution tables to obtain approximate binomial probabilities.

We may use the normal approximation to the binomial distribution when $np > 5$, $n(1 - p) > 5$, assuming of course that Table D.1 cannot be used. The conditions on n and p are more likely to be met if n is large and p is not close to 0 or 1.

Example

Suppose that one person in six is left-handed. If a class contains 40 students, what is the probability that 10 or more will be left-handed? Assuming that the four conditions for the binomial apply, this is a binomial problem with $n = 40$, $p = 1/6$, so $np = 6.67$ and $n(1 - p) = 33.33$, and hence the conditions for using the normal approximation to the binomial distribution are satisfied. We may therefore treat the variable 'number of left-handed players in a sample of 40' as though it were normally distributed with:

$$\mu = np = 6.67, \quad \sigma = \sqrt{(np(1 - p))} = 2.36$$

This distribution is shown in Fig. 7.7. Before we use Table D.3(a), we must apply a **continuity correction** of 0.5 since the 'number of left-handed players' is a discrete variable while the normal distribution is continuous. Since '10 or more on a discrete scale' is equivalent to 'more than 9.5 on a continuous scale', we use Table D.3(a) for $x = 9.5$. So:

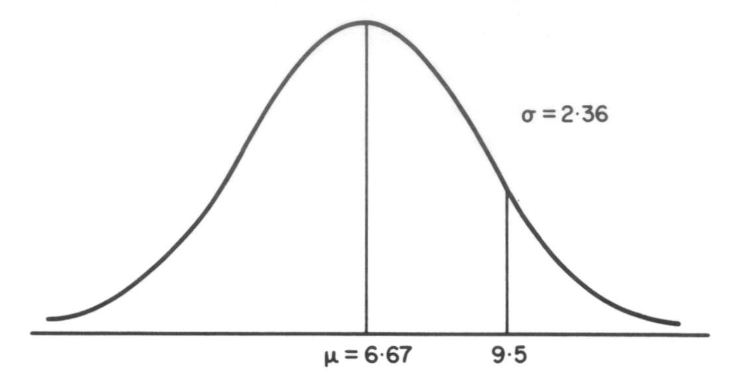

Fig. 7.7 A normal distribution with $\mu = 6.67$, $\sigma = 2.36$.

$$z = \frac{(9.5 - 6.67)}{2.36} = 1.2$$

giving an area to the left of 9.5 of 0.8849, using Table D.3(a). Hence the probability of 10 or more left-handed students in 40 is $1 - 0.8849 = 0.1151$, or 11.5%.

7.8 SUMMARY

The normal and the rectangular distributions are two standard types of continuous probability distribution. The normal distribution is the most important in statistics because it arises when a number of factors exert small positive or negative effects on the value of a variable, and because it is extremely useful in the theory of statistical inference.

Probabilities and percentages for the normal distribution may be obtained using tables or Minitab when we have numerical values for μ and σ, and for the rectangular distribution by calculating the areas of rectangles. The total area under any continuous distribution curve is 1.

The normal distribution tables may also be used to obtain approximate binomial probabilities if $np > 5$ and $n(1 - p) > 5$, and where it is not possible to use (binomial) Table D.1.

WORKSHEET 7: THE NORMAL AND RECTANGULAR DISTRIBUTIONS

1. For the normal distribution, what do μ and σ stand for?
2. Is a normal distribution ever skew?
3. For the binomial and Poisson distributions the probabilities sum

to 1. What is the equivalent property for the normal and rectangular distributions?

4. The weights of 5p oranges are normally distributed with mean 70 g and standard deviation 3 g. What percentage of these oranges weigh:

 (a) over 75 g, (b) under 60 g, (c) between 60 and 75 g?

 Given a random sample of 50 5p oranges, how many of them would you expect to have weights in categories (a), (b) and (c)?
 Suppose that the oranges are kept for a week and each loses 5 g in weight. Rework this question.

5. A fruit grower grades and prices oranges according to their diameter:

Diameter (cm)	*Price per orange (p)*
Below 5	4
Above 5 and below 6	5
Above 6 and below 7	6
Above 7 and below 8	7
Above 8 and below 9	8
Above 9 and below 10	9
Above 10	10

 Assuming that the diameter is normally distributed with a mean of 7.5 cm and standard deviation 1 cm, what percentages of oranges will be priced at each of the above prices? Given 10 000 oranges: (a) what is their total price, (b) what is their mean price?

6. A machine produces components whose thicknesses are normally distributed with a mean of 0.4 cm and standard deviation 0.02 cm. Components are rejected if they have a thickness outside the range 0.38 to 0.41 cm. What percentage are rejected? Show that the percentage rejected will be reduced to its smallest value if the mean is reduced to 0.395 cm, assuming the standard deviation remains unchanged.

7. Guests at a large hotel stay for a mean of 9 days with a standard deviation of 2.4 days. Among 1000 guests how many can be expected to stay: (a) less than 7 days, (b) more than 14 days, (c) between 7 and 14 days? Assume that length of stay is normally distributed.

8. It has been found that the annual rainfall in a town has a normal distribution, because of the varying pattern of depressions and anticyclones. If the mean annual rainfall is 65 cm, and in 15% of years the rainfall is more than 85 cm, what is the standard deviation of

annual rainfall? What percentage of years will have a rainfall below 50 cm?

9. In the catering industry the wages of a certain grade of part-time kitchen staff are normally distributed with a standard deviation of £4. If 20% of staff earn less than £30 a week, what is the mean wage? What percentage of staff earn more than £50 a week?

10. Sandstone specimens contain varying percentages of void space. The mean percentage is 15%, the standard deviation is 3%. Assuming that the percentage of void space is normally distributed, what proportion of specimens have a percentage of void space: (a) below 15%, (b) below 20%, (c) below 25%?

11. The height of adult males is normally distributed with a mean of 172 cm and standard deviation of 8 cm. If 99% of adult males exceed a certain height, what is this height?

12. A machine which automatically packs potatoes into bags is known to operate with a mean of 25 kg and a standard deviation of 0.5 kg. Assuming a normal distribution, what percentage of bags weigh: (a) more than 25 kg, (b) between 24 and 26 kg?

    To what new target mean weight should the machine be set so that 95% weigh more than 25 kg? In this case what weight would be exceeded by 0.1% of bags?

13. Using tables, show that, for any normal distribution with mean μ and standard deviation σ:
    (a) 68.26% of the area lies between $(\mu - \sigma)$ and $(\mu + \sigma)$.
    (b) 95.45% of the area lies between $(\mu - 2\sigma)$ and $(\mu + 2\sigma)$.
    (c) 99.73% of the area lies between $(\mu - 3\sigma)$ and $(\mu + 3\sigma)$.
    (d) 5% of the area lies outside the interval from $(\mu - 1.96\sigma)$ to $(\mu + 1.96\sigma)$.
    (e) 5% of the area lies above $(\mu + 1.645\sigma)$.

14. A haulage firm has 60 lorries. The probability that a lorry is available for business on any given day is 0.8. Find the probability that on any given day:
    (a) 50 or more lorries are available.
    (b) Exactly 50 lorries are available.
    (c) Fewer than 50 lorries are available.

15. An office worker commuting from Oxford to London each day keeps a record of how late his train arrives in London. He concludes that the train is equally likely to arrive at any time between 10 minutes early and 20 minutes late. If the train is more than 15 minutes late the office worker will be late for work. How often will this happen?

Repeat this question assuming now that the number of minutes late is normally distributed with $\mu = 5$ minutes, $\sigma = 7.5$ minutes. Represent both distributions in a sketch, indicating the answers as areas in the sketch.

Samples and populations

> We should extend our views far
> beyond the narrow bounds of
> a parish, we should include large
> groups of mankind.

8.1 INTRODUCTION

In the remaining chapters we shall be concerned mainly with statistical inference, by which we mean drawing conclusions from sample data about the larger populations from which the samples are drawn. Although we have met the words 'sample' and 'population' in earlier chapters, they were not defined. Here are the definitions:

A **population** is the whole set of measurements or counts about which we want to draw a conclusion. If we are interested in only one variable we call the population univariate; for example, the heights of all female students in higher education form a univariate population. Notice that a population is a set of measurements, not the individuals or objects on which the measurements or counts are made.

A **sample** is a subset of a population, a set of some of the measurements or counts which comprise the population.

8.2 REASONS FOR SAMPLING

The first and most obvious reason for sampling is to save time, money and effort. For example, in opinion polls before a general election it would not be practicable to ask the opinion of the whole electorate on how it intends to vote.

The second and less obvious reason for sampling is that, even though we have only part of all the information about the population, nevertheless the sample data can be useful in drawing conclusions about the population, provided that we use an appropriate **sampling method** (section 8.3) and choose an appropriate **sample size** (section 8.4).

The third reason for sampling applies to the special case where the act of measuring the variable destroys the 'individual', such as in the destructive testing of explosives. Clearly, testing a whole batch of explosives would be ridiculous.

8.3 SAMPLING METHODS

There are many ways of selecting a sample from a population, but the most important is **random sampling**. Indeed, the methods of statistical inference used in this book apply only to cases in which the sampling method is random. Nevertheless, there are other sampling methods which are worth mentioning for reasons stated below.

A **random sample** is defined as one for which each measurement or count in the population has the same chance (probability) of being selected. A sample selected so that the probabilities are not the same for each measurement or count is said to be a **biased** sample. For the population of female student heights, selecting only those who were members of netball teams might result in a biased sample since they would tend to be taller than the average female student. Random sampling requires that we can identify all the individuals or objects which comprise the population, and that each measurement or count to be included in the sample is chosen using some method which ensures equal probability, for example by the use of random number tables, such as Table D.4 of Appendix D.

Example

For the student height example, suppose that each student is assigned a unique five-digit number (assuming that there are fewer than 100 000 female students in total). Starting anywhere in Table D.4, read off 5 consecutive digits by moving from one digit to the next in any direction – up, down, left, right or diagonally. This procedure could result in the number 61978, say. This number identifies the first student, and her height is the first sample value to be observed. If there is no student with this number, we ignore it and select another random number. We carry on until the required number of observations, called the sample size, has been reached.

Random numbers are also available on most scientific calculators and on computers.

Where the individuals are objects which are fixed in a given location, the method of random sampling can also be used. There are many examples in geography, geology and environmental biology where such samples are required. One method of obtaining a random sample is to

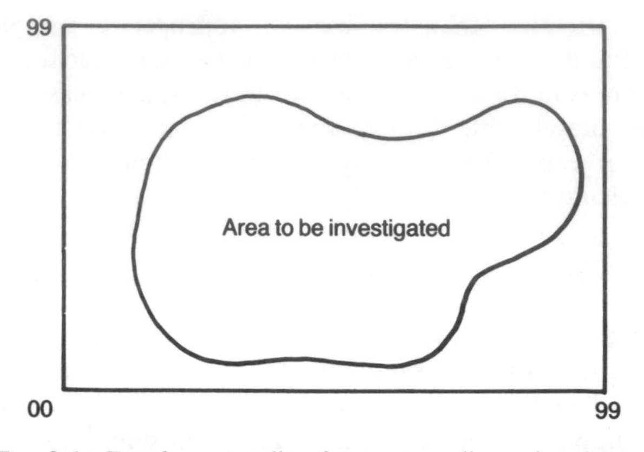

Fig. 8.1 Random sampling from a two-dimensional area.

overlay a plan of the area to be investigated with a rectangular grid which includes all points of interest in the area (Fig. 8.1).

From one corner of the rectangle one side is labelled with the 100 values, 00 to 99, spaced at equal intervals. The other side is labelled in the same way. By selecting two 2-digit random numbers, the coordinates of a randomly selected point within the rectangle are determined. If this point falls outside the area of interest it is ignored and two more random numbers are selected. The procedure is repeated until the required number of points has been selected.

Systematic sampling may be used to cut down the time taken in selecting a random sample. In the student height example, suppose that we require a 1% (1 in 100) sample from 80 000 students. We could select one number in the range 1 to 100 and derive all other student numbers to be included in the sample by adding 100, 200, 300, and so on, giving a sample size of 800 in all. Systematic sampling is suitable in that it provides a quasi-random sample so long as there is no periodicity in the population list or geographical arrangement which coincides with the selected numbers. For example, if we were selecting from a population of houses in an American city where the streets are laid out in a grid formation, then selecting every 100th house could result in always choosing a corner house on an intersection of two roads, possibly giving a biased sample.

Stratified sampling may be used where it is known that the individuals or objects to be sampled provide not one population but a number of distinct sub-populations or **strata**. These strata may have quite different distributions for the variable of interest.

Example

For the population of the heights of female students we may wish to separate first-, second- and third-year students, for example. If 40% of students are first-years, 30% are second-years, and 30% are third-years, we could take a 1 in 100 random sample from each of these three strata to provide an overall 1% sample.

Other methods include quota sampling, cluster sampling, multi-stage and sequential sampling. Each method has its own special application area and will not be discussed here.

8.4 SAMPLE SIZE

The most common question asked by every investigator who wishes to collect and analyse data is, 'How much data should I collect?'. We refer to the number of observations to be included in the sample as the **sample size**, so the investigator should be asking, 'What sample size should I choose?'. If we wish to give a slightly more sophisticated answer than, for example, 'a sample size of 20 seems a bit too small, but 30 sounds about right', we can use the arguments of the following example.

Example

Suppose we wish to estimate the (population) mean height, μ, of female students in higher education. The required sample size should depend on two factors:

1. The precision we require for the estimate, which the investigator must specify, knowing that the more precision he requires, the larger the sample size must be.
2. The variability of height, as measured by its standard deviation. We would think it reasonable to believe that the larger the standard deviation of height, the larger will be the required sample size. The standard deviation can only be determined when we have some data. But since we are trying to decide how much data to collect, we are in a chicken-and-egg situation.

We will return to a discussion of sample size when we have discussed the ideas of confidence intervals in the next chapter.

8.5 SAMPLING DISTRIBUTION OF THE SAMPLE MEAN

This is not a book about the mathematical theory of statistics, but the theory which we now discuss is essential for a more complete understanding of the inferential methods to be discussed later, particularly those concerning confidence intervals (in the next chapter).

Suppose that we are sampling from a population of measurements which has a mean, μ, and standard deviation, σ. We will suppose that the

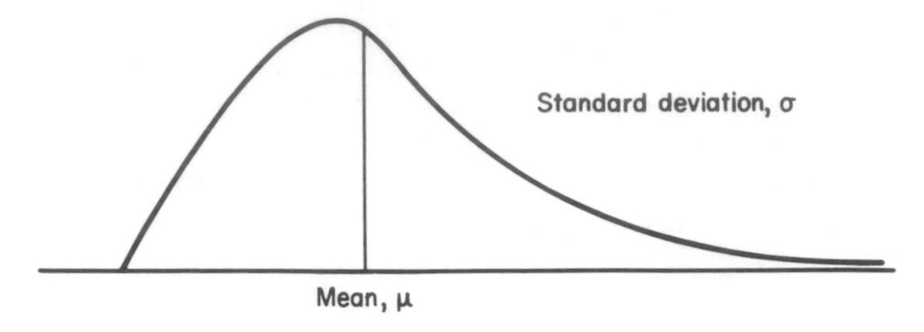

Fig. 8.2 A continuous probability distribution.

measurement is of a continuous variable so that its probability distribution is continuous but not necessarily normal (Fig. 8.2).

Suppose we select a random sample of size n from this 'parent' population, and calculate $\bar{x}$, the sample mean (using the methods of section 4.2). If we continue this procedure of taking random samples of size n and calculating the sample mean on each occasion, we will eventually have a large number of sample means. Since the random samples are very unlikely to consist of exactly the same observations, these sample means will vary and form a new population with a distribution which is called the **sampling distribution of the sample mean**.

This distribution will have a mean and standard deviation which we will denote by $\mu_{\bar{x}}$ and $\sigma_{\bar{x}}$ to distinguish them from the μ and σ of the parent population. The sampling distribution will obviously have a shape. The question of interest is, 'What is the connection between the mean, standard deviation and shape of the distribution of the parent population and the mean, standard deviation and shape of the sampling distribution of the sample mean?'.

The answer to this question is supplied now by some of the most important results in statistical theory which we quote (without proof!) in three parts:

1. $\mu_{\bar{x}} = \mu$
2. $\sigma_{\bar{x}} = \sigma/\sqrt{n}$
3. If n is large, the distribution of the sample mean is approximately normal, irrespective of the shape of the distribution of the parent population. (If, however, the 'parent' is normal, then the distribution of the sample mean is also normal for all values of n.)

The beauty of mathematics (if non-mathematicians will allow me to eulogize for just one sentence) is that so much can be stated in so few symbols. However, we can also use a picture (Fig. 8.3) and some words to explain the three parts of this theory.

In words the three parts of the theory state that, if lots of random samples of the same size, n, are taken from a parent distribution, then:

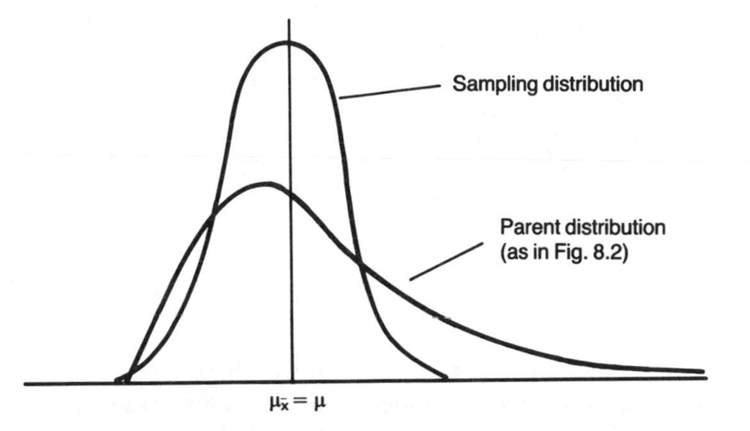

Fig. 8.3 A 'parent' distribution and the sampling distribution of the sample mean.

1. The mean of the sample means will equal the mean of the parent.
2. The standard deviation of the sample means will be smaller than the standard deviation of the parent by a factor of $\sqrt{n}$. For example, if $n = 4$, the standard deviation of the sample means will be half the standard deviation of the parent (since $\sqrt{4} = 2$).
3. The shape of the sampling distribution of the sample mean is approximately normal for large sample sizes, even if the parent distribution is not normal (and the larger the sample size, the more closely is this distribution normal).

This third result is called the **central limit theorem** and is an example of one of the two reasons (section 7.2) for the importance of the normal distribution in statistics.

In the next chapter we will use the three results quoted above to make estimates of the mean of a population from sample data (these estimates are called **confidence intervals**).

8.6 SIMULATION OF THE SAMPLING DISTRIBUTION OF THE SAMPLE MEAN USING MINITAB

We can decide whether the three-part theory quoted in section 8.5 seems reasonable by simulating samples from a known parent population and looking at aspects of the distribution of the sample mean. An example is given in Table 8.1.

The RANDOM command and the sub-command which follows put 200 randomly selected observations from a normal distribution with a mean

Table 8.1 Distribution of sample means of samples of size 9 taken from a $N(163, 6^2)$ distribution

```
MTB> RANDOM  200  C1-C9;
SUBC> NORMAL  MU = 163, SIGMA = 6.
MTB> RMEAN  C1-C9  INTO  C10
MTB> PRINT  C10
MTB> DESCRIBE  C10
MTB> HISTOGRAM  C10
```

of 163 and a standard deviation of 6 into each of columns 1 to 9. We therefore have 200 rows of observations in our worksheet, each row consisting of nine observations. The RMEAN command calculates the mean of each row, i.e. sample, of size 9 and puts the answer in column 10. Column 10 is thus a simulation of 200 observations from the distribution of the sample mean, for samples of size 9. The theoretical values for the mean and standard deviation of this distribution are 163 and $6/\sqrt{9} = 2$, respectively, and we can see how close we get to these by the output following the DESCRIBE command. We can also check visually to see whether the sampling distribution seems normal by using the HISTOGRAM command.

The reader is urged to try:

1. varying the number of samples selected (200 in this example) for the same 'parent';
2. varying the sample size (9 in the example) for the same 'parent';
3. a non-normal 'parent', for example a Poisson distribution with a mean of 0.5.

8.7 SUMMARY

A population is the whole set of measurements or counts about which we want to draw a conclusion, and a sample is a subset of a population. Conclusions may be drawn from sample data if an appropriate sampling method and sample size are chosen. Random sampling is important because the theory of how to make inferences about populations from randomly sampled data is well developed. Systematic sampling may sometimes be used instead of random sampling, and stratified sampling is appropriate when the population is, in reality, made up of a number of sub-populations. The required sample size depends on the precision required in the estimate of the population parameter, and on the variability in the population.

The theory of inference depends on the distribution of sample statistics, such as the sample mean. The theory relates the characteristics of the sampling distribution to those of the parent population.

It is possible to simulate the sampling distribution of the sample mean using Minitab.

WORKSHEET 8: SAMPLES AND POPULATIONS

1. What is: (a) A population? (b) A sample? (c) A random sample? (d) A biased sample? (e) A census?

2. Why and how are random samples taken? Think of an example in your main subject area of interest.

3. What is wrong with each of the following methods of sampling? In each case think of a better method.
   (a) In a laboratory experiment with mice, five mice were selected by the investigator by plunging a hand into a cage containing 20 mice and catching them one at a time.
   (b) In a survey to obtain adults' views on unemployment, people were stopped by the investigator as they came out of (i) a travel agent, (ii) a food supermarket, (iii) a job centre.
   (c) In a survey to decide whether two species of plant tended to grow close together in a large meadow, the investigator went into the meadow and randomly threw a quadrat over his left shoulder a number of times, each time noting the presence or absence of the species (a quadrat is, typically, a metal or plastic frame 1 metre square).
   (d) A survey of the price of bed and breakfast was undertaken in a town with 40 three-star, 50 two-star and 10 one-star hotels. A sample of ten hotels was obtained by numbering the hotels from 00 to 99 and then using random number tables.
   (e) To save time in carrying out an opinion poll in a constituency, a party worker selected a random sample from the electoral register for the constituency, and telephoned those selected.
   (f) To test the efficacy of an anti-influenza vaccine in his practice, a doctor placed a notice in his surgery asking patients to volunteer to be given the vaccine.

4. Answer this question using Minitab only.

   When a die is thrown once, the discrete probability distribution for the number on the uppermost face is:

Number	1	2	3	4	5	6
Probability	1/6	1/6	1/6	1/6	1/6	1/6

The mean and standard deviation of the distribution are $\mu = 3.5$, $\sigma = 1.71$. Now simulate the following experiment: throw two dice and calculate the mean of the two scores on the uppermost face. Repeat this experiment a total of 108 times and obtain a histogram, noting that the possible values for the mean score are 1, 1.5, 2.0, . . . , 6.0. Calculate the mean and standard deviation of this distribution. (Assuming these are reasonably good estimates of μ and σ for samples of size $n = 2$, you should find that $\mu_{\bar{x}} = \mu = 3.5$ and $\sigma_{\bar{x}} = \sigma/\sqrt{2} = 1.21$, approximately. You should also find that the shape of the distribution is approximately triangular, which is a little more normal than the flat shape of the 'parent'.)

Repeat this simulated experiment for three dice.

Confidence interval estimation

9.1 INTRODUCTION

Consider the random sample of 27 female student heights from column 3 of Appendix A. These heights are represented graphically in part of Fig. 3.11, reproduced here as Fig. 9.1.

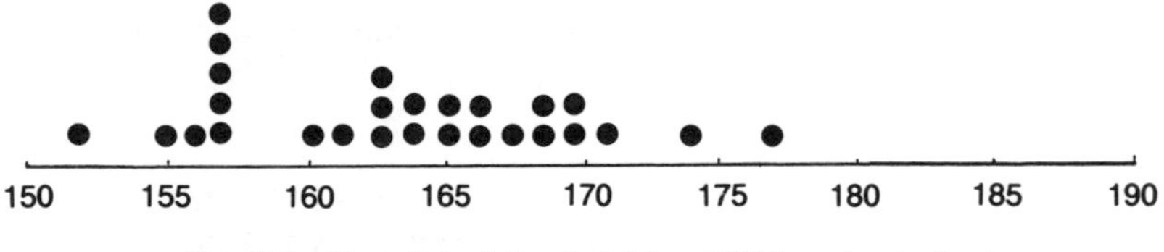

Fig. 9.1 Dotplot of the heights of 27 female students.

By calculation we find that the sample mean height is $\bar{x} = 163.4$ cm. What is our estimate of the population mean height μ (i.e. of all female students in higher education)? The obvious answer is 163.4 cm if we require a single-value estimate (often referred to as a **point** estimate). However, since our estimate is based on only a sample of the population of heights, we might want to be more guarded and give some idea of the precision of the estimate by adding an **error term** which might, for example, lead to a statement that the estimate is 163.4 ± 1, meaning that the population mean height, μ, lies between 162.4 and 164.4 cm. Such an estimate is referred to as an **interval** estimate. The size of the error term will depend on three factors:

1. The sample size, n. The larger the sample size, the better the estimate, the smaller the error term and the greater the precision.
2. The variability in height (as measured by the standard deviation). The larger the standard deviation, the poorer the estimate, the larger the error term and the smaller the precision.

3. The level of confidence we wish to have that the population mean height does in fact lie within the specified interval. This is such an important concept that we will devote the next section to it.

9.2 95% CONFIDENCE INTERVALS

For the student height example, the population mean height, μ, has a fixed numerical value at any given time. We do not know this value, but by taking a random sample of 27 heights we want to specify an interval within which we want to be reasonably confident that this fixed value lies.

Suppose we decide that nothing less than 100% confidence will suffice, implying absolute certainty. Unfortunately, theory indicates that the 100% confidence interval is so wide that it is useless for all practical purposes. Typically, investigators choose a 95% **confidence level** and calculate a 95% **confidence interval** for the population mean, μ, using formulae we shall introduce in the next sections. For the moment it is important to understand what a confidence level of 95% means. It means that on 95% of occasions when such intervals are calculated the population mean will actually fall inside the calculated interval, but on 5% of occasions it will fall outside the interval. In a particular case, however, we will not know whether the mean has been successfully 'captured' within the calculated interval.

9.3 CALCULATING A 95% CONFIDENCE INTERVAL FOR THE MEAN, μ, OF A POPULATION: LARGE SAMPLE SIZE, n

Having discussed some concepts, we now discuss methods of actually calculating a 95% confidence interval for the population mean, μ.

In section 8.5, we stated that the sample mean has a distribution with mean μ and standard deviation $\sigma/\sqrt{n}$, and is approximately normal if n is large (Fig. 9.2).

Recall that μ and σ were the mean and standard deviation of the 'parent' distribution. It follows from the theory of the normal distribution that 95% of the values of the sample mean lie within 1.96 standard deviations of the mean of the distribution of sample means (refer to section 7.2 if necessary).

So 95% of the values of the sample means lie in the range $\mu - 1.96\sigma/\sqrt{n}$ to $\mu + 1.96\sigma/\sqrt{n}$. We can write this as a probability statement:

$$P(\mu - 1.96\sigma/\sqrt{n} < \bar{x} < \mu + 1.96\sigma/\sqrt{n}) = 0.95$$

This statement can be rearranged to state:

$$P(\bar{x} - 1.96\sigma/\sqrt{n} < \mu < \bar{x} + 1.96\sigma/\sqrt{n}) = 0.95$$

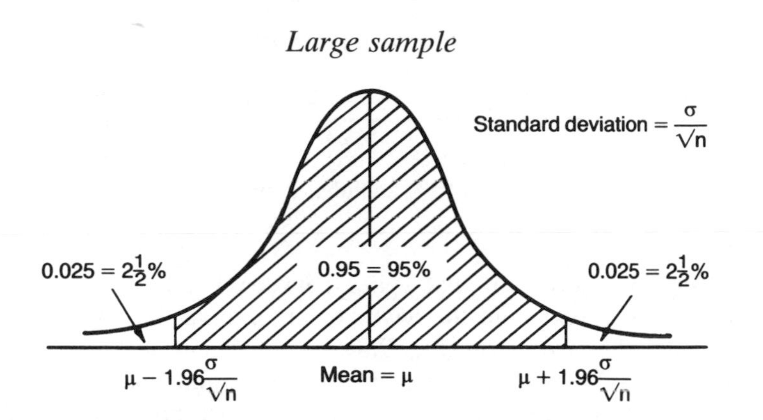

Fig. 9.2 Distribution of the sample mean $\bar{x}$, if n is large.

If we now use the fact that n is large, we could replace σ by the sample standard deviation, s, which we calculate from sample data, and obtain the following statement, which is now approximately true:

$$P(\bar{x} - 1.96s/\sqrt{n} < \mu < \bar{x} + 1.96s/\sqrt{n}) = 0.95$$

The importance of this result is that we can calculate $\bar{x} - 1.96s/\sqrt{n}$ and $\bar{x} + 1.96s/\sqrt{n}$ from our sample data. These values are called the 95% **confidence limits** for μ.

The interval $\bar{x} - 1.96s/\sqrt{n}$ to $\bar{x} + 1.96s/\sqrt{n}$ is called the 95% **confidence interval** for μ. The **error** term (referred to in section 9.1) is $1.96s/\sqrt{n}$. Note that we can only use the formulae above when n is large (greater than 30, say).

(Some texts discuss the case where σ, the population standard deviation, is known (as an exact value). This author's view is that this is unrealistic since, if we know σ, we must have all the measurements in the population and hence we also know μ, the population mean, and so there is no need to estimate it.)

Example

Suppose that, from a random sample of 27 heights, we find that $\bar{x}$ = 163.4 cm, and s = 6.1 cm (these summary statistics apply to the data referred to in section 9.1). Then a 95% confidence interval for the mean height, μ, is:

$$163.4 - \frac{1.96 \times 6.1}{\sqrt{27}} \quad \text{to} \quad 163.4 + \frac{1.96 \times 6.1}{\sqrt{27}}$$

$$163.4 - 2.3 \quad \text{to} \quad 163.4 + 2.3$$

$$161.1 \quad \text{to} \quad 165.7$$

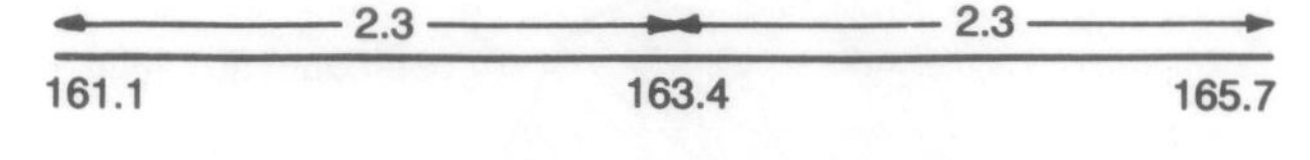

Fig. 9.3 A 95% confidence interval for μ.

So we are 95% confident that μ lies between 161.1 and 165.7 cm, the 'error' term is 2.3 cm, and the width of the interval is $165.7 - 161.1 = 4.6$ cm, twice the error term (Fig. 9.3).

The formula for the error term is $1.96s/\sqrt{n}$, and this can be used to support the intuitive arguments of section 9.1 concerning the three factors which affect the 'error' term:

1. Clearly as n increases, $1.96s/\sqrt{n}$ decreases.
2. The larger the variability, the larger s will be, so $1.96s/\sqrt{n}$ will be larger.
3. We use the factor 1.96 in the error term for 95% confidence. For greater confidence we would need to move further away from the mean (see Fig. 9.2), so the factor 1.96 would increase and so would the error term. So the greater the confidence level, the greater the error term, and the wider the interval.

9.4 CALCULATING A 95% CONFIDENCE INTERVAL FOR THE MEAN, μ, OF A POPULATION: SMALL SAMPLE SIZE, n

The formula of the previous section referred to cases where the sample size is large; this can be taken to mean 'greater than 30' as a rough guide. If we assume that the variable of interest, x, is approximately normally distributed and $n > 1$, a 95% confidence interval for μ is given by:

$$\bar{x} - \frac{ts}{\sqrt{n}} \quad \text{to} \quad \bar{x} + \frac{ts}{\sqrt{n}},$$

sometimes written as

$$\bar{x} \pm \frac{ts}{\sqrt{n}}.$$

The value of t is obtained from Table D.5 of Appendix D and depends on two factors:

1. The confidence level, which determines the value of α to be entered in Table D.5. For example, for 95% confidence, $\alpha = (1 - 0.95)/2 = 0.025$.
2. The sample size, n, which determines the number of degrees of

freedom, ν, to be entered in Table D.5. The general idea of what is meant by the term 'degrees of freedom' is discussed in section 9.7. In using the formula above, $\nu = (n - 1)$.

Note also that the 'error' term in the formula is $ts/\sqrt{n}$.

We stated above that we are making the assumption that x is approximately normally distributed. However, this assumption is less critical the larger the value of n.

The difficulty with this assumption arises when n is small, say below 15. How can you tell whether a small sample has been taken from a normal distribution if all you have are a few observed values? You can, I suppose, draw a dotplot and look for two characteristics of any normal distribution, namely symmetry and bunching in the middle. Clearly the following is approximately normal:

● ● ● ●●●●●●●●●● ● ● ●

But what about this? It exhibits symmetry, but not bunching:

● ● ● ● ● ●● ● ● ● ● ●● ●●

And what about this? It exhibits bunching, but not symmetry:

● ●●●●●●●●●● ● ● ● ●

Here is a suggestion! Use the Minitab simulation method of section 7.5 to take a sample of 15 observations from any normal distribution, for example $N(0,1)$, and draw a dotplot. Repeat this for several other random samples of the same size from the same distribution and get a 'feel' for what small random samples taken from a normal distribution look like. Figure 9.4 shows five I did earlier.

Example

Suppose we take a sample of only nine female student heights and obtain the following observations:

163, 157, 160, 168, 155, 168, 164, 157, 169

From these data, $\bar{x} = 162.3$, $s = 5.3$. A dotplot for the data is given in Fig. 9.5.

I would be quite happy to assume approximate normality here, not only because the dotplot looks reasonable but also because the heights here are of a sample from a population of nominally identical individuals. Nevertheless, this is a subjective decision. If you are not happy to assume normality here, there are two things you could do:

1. Carry out a formal test of normality, as discussed in section 15.5.

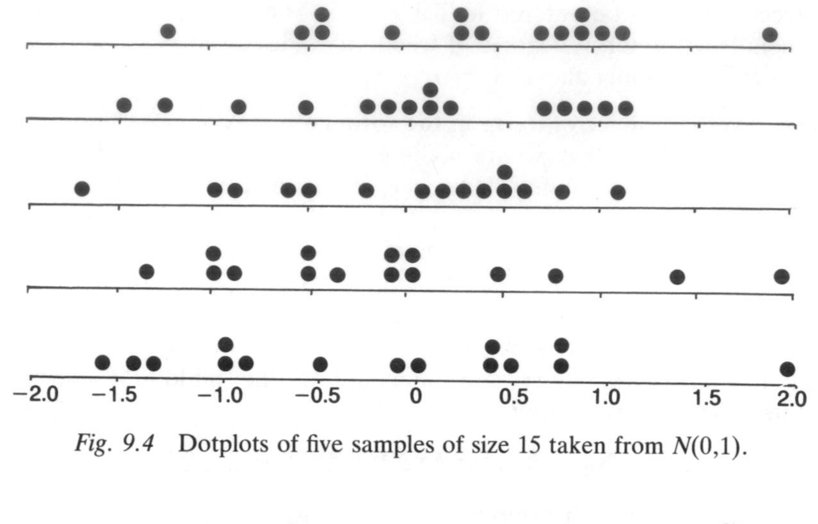

Fig. 9.4 Dotplots of five samples of size 15 taken from $N(0,1)$.

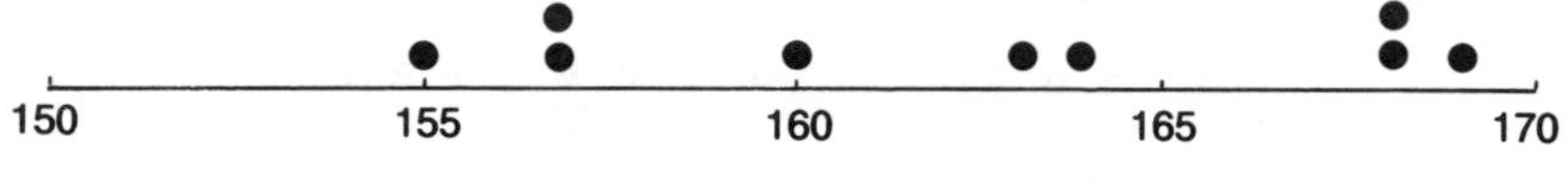

Fig. 9.5 Dotplot of the heights of nine female students.

2. Take a larger sample, both to give a better plot and also to reduce the importance of the assumption of normality.

We now wish to find a value for t. For 95% confidence, $\alpha = 0.025$. For $n = 9$, $\nu = 9 - 1 = 8$. Hence from Table D.5, $t = 2.306$. Thus a 95% confidence interval for μ is:

$$162.1 - \frac{2.306 \times 5.3}{\sqrt{9}} \quad \text{to} \quad 162.1 + \frac{2.306 \times 5.3}{\sqrt{9}}$$

$$162.1 - 4.1 \quad \text{to} \quad 162.1 + 4.1$$

$$158.0 \quad \text{to} \quad 166.2$$

Based on a sample of 9 observations, we are 95% confident that μ lies between 158.0 and 166.2. The width of the interval has grown from 4.6 for $n = 27$ to 8.2 for $n = 9$, i.e. nearly doubled.

Note that if we use t tables for large values of n, then for 95% confidence, and hence $\alpha = 0.025$, the t value is 1.96 and we have the formula we used in section 9.3. We may therefore use the formula $\bar{x} \pm ts/\sqrt{n}$ for all $n \geqslant 2$, and stop wondering what 'n must be large' means. However, we must still be able to justify approximate normality if n is small.

Minitab also can be used to obtain a 95% confidence interval for μ, using the command TINTERVAL, as shown in Table 9.1.

Table 9.1 A 95% confidence interval for μ, the Minitab commands and output

```
MTB> SET  C1
DATA> 163  157  160  168  155  168  164  157  169
DATA> END
MTB> TINTERVAL  C1
         N       MEAN      STDEV      SEMEAN      95.0 PERCENT C.I.
C1       9      162.3        5.3         1.8       (158.0, 166.2)
```

9.5 THE t DISTRIBUTION

This continuous probability distribution was first studied by W.S. Gosset, who published his results under the pseudonym of Student, which is why the distribution is often referred to as Student's t distribution. It arises when we consider taking a large number of random samples of the same size, n, from a normal distribution with known mean, μ. Then the probability distribution of the statistic:

$$t = \frac{\bar{x} - \mu}{s/\sqrt{n}}$$

may be plotted. It will be symmetrical and unimodal. For different values of n, different distributions will be obtained; for large n the t distribution approaches the standardized normal distribution, while for small n the t distribution is flatter and has higher tails than the normal distribution (Fig. 9.6).

9.6 THE CHOICE OF SAMPLE SIZE WHEN ESTIMATING THE MEAN OF A POPULATION

In section 8.4 we discussed the choice of sample size, n, but deferred deciding how to calculate how large it should be until this chapter.

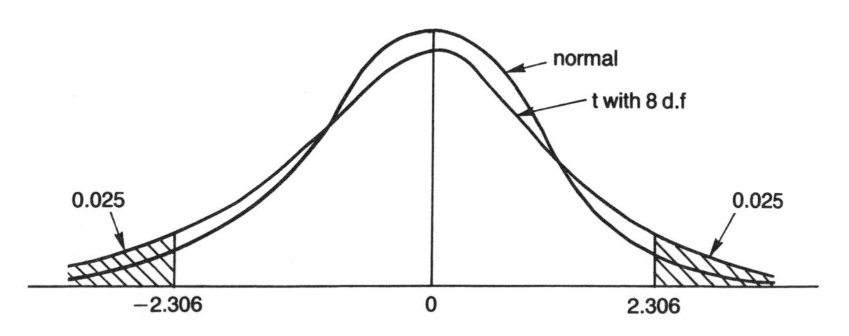

Fig. 9.6 Comparison of the shapes of a normal distribution and a t distribution with $\nu = 8$ degrees of freedom.

Instead we concentrated on the factors affecting the choice of n for the case of estimating μ, the mean of a population. These factors were:

1. The precision with which the population mean is to be estimated, and we can now state this precision in terms of the 'error' term in the formula for the confidence interval for μ.
2. The variability of the measurements, and we noted a chicken-and-egg situation of needing to know the variability before we had any sample data. In order to overcome this difficulty we can carry out a pilot experiment and obtain at least a rough estimate of the standard deviation.

Example

Suppose we specify a precision in terms of an 'error' term of 1 for 95% confidence. Then $ts/\sqrt{n} = 1$, where t is found fom Table D.5 for $\alpha = 0.025$, but n and hence $\nu = (n - 1)$ are unknown.

Suppose further that we also have a rough estimate that $s = 10$ from a small pilot experiment. Then:

$$\frac{t \times 10}{\sqrt{n}} = 1$$

How can we solve this equation to find n, since t depends on n? The trick is to assume that n is large, and note that, for $\alpha = 0.025$, t is roughly 2 for large values of n. Now we can solve

$$\frac{2 \times 10}{\sqrt{n}} = 1$$

to obtain $n = 20^2 = 400$ as the required sample size. (We were correct in assuming n would be large.)

9.7 DEGREES OF FREEDOM

There are two approaches which you, the reader, can take to the concept of degrees of freedom. The 'cookbook' approach is to know where to find the formula for calculating degrees of freedom for each application covered (and there are quite a few in the remaining chapters). The more mature approach is to try to understand the general principle behind all the formulae for degrees of freedom, which is as follows. The number of degrees of freedom equals the number of independent values used in calculating a statistic, minus the number of restrictions placed on the data.

Example

Why do we use $(n - 1)$ degrees of freedom when we look up t in Table D.5 in the calculation of 95% confidence limits for μ? The answer is that in the formula $\bar{x} \pm ts/\sqrt{n}$, we calculate s using

$$s = \sqrt{\frac{\Sigma x^2 - \frac{(\Sigma x)^2}{n}}{n - 1}}$$

An alternative form of this formula (see section 4.7) is:

$$s = \sqrt{\frac{\Sigma(x - \bar{x})^2}{n - 1}} = \sqrt{\frac{(x_1 - \bar{x})^2 + (x_2 - \bar{x})^2 + \cdots (x_n - \bar{x})^2}{n - 1}}$$

Only $(n - 1)$ of these differences $(x_1 - \bar{x})$, $(x_2 - \bar{x})$, etc., are independent, since there is a restriction that the sum $\Sigma(x - \bar{x}) = 0$ (we saw examples of this in Worksheet 2, Questions 4 and 5).

9.8 95% CONFIDENCE INTERVAL FOR A BINOMIAL PROBABILITY

All the discussion so far in this chapter has been concerned with confidence intervals for the mean, μ, of a population. If our sample data are from a binomial experiment for which we do not know the value of p, the probability of success in each trial (in other words, the proportion of successes in a large number of trials), we can use our sample data to calculate a 95% confidence interval for p. Thus if we observed x successes in the n trials of a binomial experiment, then a 95% confidence interval for p is

$$\frac{x}{n} \pm 1.96\sqrt{\frac{\frac{x}{n}\left(1 - \frac{x}{n}\right)}{n}}$$

provided $x > 5$ and $n - x > 5$. These two conditions are the equivalent of $np > 5$ and $n(1 - p) > 5$ for the normal approximation to the binomial (section 7.7), where the unknown p is replaced by its point estimator, x/n. Note also that in using this formula, the four conditions for the binomial must apply (see section 6.3).

Example

Of a random sample of 200 voters taking part in an opinion poll, 110 said they would vote for party A, the other 90 said they would vote for other parties. What proportion of the total electorate will vote for party A?

If we regard 'voting for A' as a 'success', then $x = 110$, $n = 200$. The conditions $x > 5$ and $n - x > 5$ are satisfied, so a 95% confidence interval for p is:

$$\frac{110}{200} \pm 1.96\sqrt{\frac{\frac{110}{200}\left(1 - \frac{110}{200}\right)}{200}}$$

$$\text{i.e.} \quad 0.55 \quad \pm \quad 0.07$$

$$\text{or} \quad 0.48 \quad \text{to} \quad 0.62$$

We can be 95% confident that the proportion who will vote for party A is between 0.48 (48%) and 0.62 (62%).

9.9 THE CHOICE OF SAMPLE SIZE WHEN ESTIMATING A BINOMIAL PROBABILITY

In the example of the previous section the width of the confidence interval is quite large. If we wished to reduce the width by reducing the error term, we would need to increase the sample size.

Example

If we wished to estimate the proportion to within (an error term of) 0.02 for 95% confidence, the new sample, n, could be found by solving:

$$1.96\sqrt{\frac{\frac{110}{200}\left(1 - \frac{110}{200}\right)}{n}} = 0.02$$

That is

$$n = \frac{1.96^2 \times 0.55 \times 0.45}{0.02^2} = 2377$$

We need a sample of nearly 2500. Notice how we again used the result of a pilot survey (of 200 voters) as in section 9.6.

9.10 95% CONFIDENCE INTERVAL FOR THE MEAN OF A POPULATION OF DIFFERENCES, 'PAIRED' SAMPLES DATA

In experimental work we are often concerned with not just one population, but with a comparison between two populations. For example, suppose that two methods of teaching children to read are to be compared. Some children are taught by a standard method (S) and some by a new method (N). In order to reduce the effect of factors other than teaching method, children are matched in pairs so that children in each pair are as similar as possible with respect to factors such as age, sex, social background and initial reading ability. One child from each pair is then randomly assigned to teaching method S and the other to method N. Suppose that after one year the children are tested for reading ability, and that the data in Table 9.2 are the test scores for 10 pairs of children.

In this example we can think of **two** populations of measurements, namely the S method scores and the N method scores. However, if our main interest is in the difference between the methods, the **one** population about which we wish to draw inferences is the **population of differences** between pairs.

The sample data in Table 9.2 are an example of **paired** samples data. The differences, d, have been calculated and put into the bottom row of Table 9.2. Then a 95% confidence interval for μ_d, the mean of the population of differences, can be calculated using

$$\bar{d} \pm \frac{ts_d}{\sqrt{n}}$$

where $\bar{d}$ is the sample mean difference, so $\bar{d} = (\Sigma d)/n$, and s_d is the sample standard deviation of differences. So

$$s_d = \sqrt{\frac{\Sigma d^2 - \frac{(\Sigma d)^2}{n}}{n - 1}}$$

In this formula, n is the number of differences (= number of pairs) and t is obtained from Table D.5 for $\alpha = 0.025$, $\nu = (n - 1)$.

In using this formula, we need to be able to assume that the differences are approximately normally distributed. This assumption is less critical the larger the value of n.

For the data in Table 9.2, a dotplot is given in Fig. 9.7. The differences do not seem to be markedly non-normal. Also for these data, $\bar{d} = 7.5$, $s = 8.48$, $n = 10$, $t = 2.262$, so a 95% confidence interval for μ_d is:

Table 9.2 Reading test scores of 10 matched pairs of children

Pair number	1	2	3	4	5	6	7	8	9	10
S method score	56	59	61	48	39	56	75	45	81	60
N method score	63	57	67	52	61	71	70	46	93	75
$d = N$ score $- S$ score	7	−2	6	4	22	15	−5	1	12	15

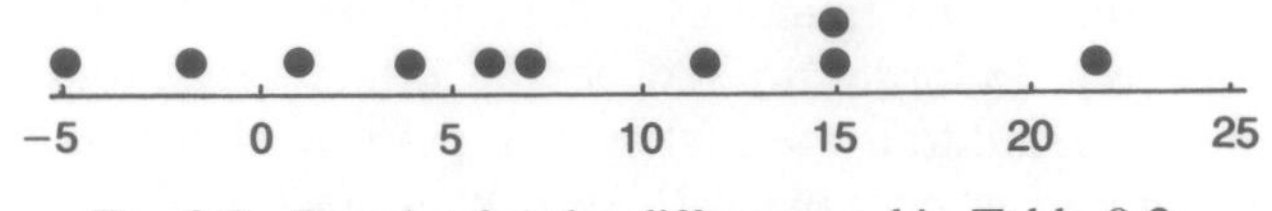

Fig. 9.7 Dotplot for the differences d in Table 9.2.

$$7.5 \pm \frac{2.262 \times 8.48}{\sqrt{10}}$$

i.e. 1.4 to 13.6.

We are 95% confident that the mean population difference in scores between the methods lies between 1.4 and 13.6, where a positive difference means that the N score is higher.

As in section 9.6, it would now be possible to decide what sample size to choose in another experiment designed to provide a more precise estimate of the mean difference in scores between the methods.

To use Minitab to obtain a 95% confidence interval for μ_d, we can either input the differences and use TINTERVAL as in Table 9.1, or we can input the two sets of scores and get Minitab to calculate the differences as in Table 9.3.

9.11 95% CONFIDENCE INTERVAL FOR THE DIFFERENCE IN THE MEANS OF TWO POPULATIONS, 'UNPAIRED' SAMPLES DATA

The example of the previous section was not, in essence, a comparison of two populations since the data were in pairs. In many other instances where two populations of measurements are concerned, the data are **unpaired**.

For example, the A-level counts of a random sample of students

Table 9.3 Confidence interval for μ_d, paired samples data

```
MTB> SET  C1
DATA> 56  59  61  48  39  56  75  45  81  60
DATA> END
MTB> SET C2
DATA> 63  57  67  52  61  71  70  46  93  75
DATA> END
MTB> LET  C3 = C2 - C1
MTB> NAME  C3  'DIFFS'
MTB> TINTERVAL  C3
```

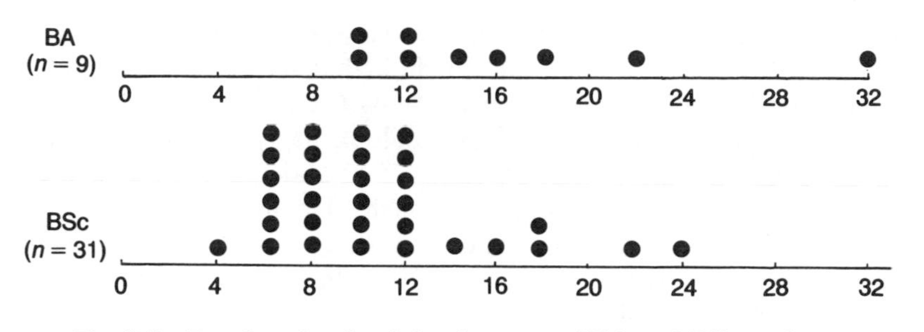

Fig. 9.8 Dotplots for the A-level counts of BA and BSc students.

studying for a BA or BSc degree were summarized graphically as shown in Fig. 9.8 from the data in column 7 of Appendix A. Calculate a 95% confidence interval for $(\mu_1 - \mu_2)$, the difference in the means of the two populations, where μ_1 is the population mean A-level count for BA students, and μ_2 is the population mean A-level count for BSc students.

Note that the data in Fig. 9.8 are unpaired in the sense that no A-level count in the first sample (BA students) is associated with any particular A-level count in the second sample (BSc students).

The formula for a 95% confidence interval for $(\mu_1 - \mu_2)$ is

$$(\bar{x}_1 - \bar{x}_2) \pm ts\sqrt{\frac{1}{n_1} + \frac{1}{n_2}}$$

where $\bar{x}_1$ is the sample mean for the first sample of size n_1; $\bar{x}_2$ is the sample mean for the second sample of size n_2; t is found from Table D.5 for $\alpha = 0.025$, $\nu = (n_1 + n_2 - 2)$ degrees of freedom; and s^2, given by

$$s^2 = \frac{(n_1 - 1)s_1^2 + (n_2 - 1)s_2^2}{n_1 + n_2 - 2}$$

is a weighted average of the two sample variances s_1^2 and s_2^2, and is called a **pooled** estimate of the common variance of the two populations (see assumption 2 below).

In using the formula above, we make the following assumptions:

1. The measurements in each population must be approximately normally distributed, this assumption being less critical the larger the values of n_1 and n_2.
2. The population standard deviations, σ_1 and σ_2, must be equal.

For the numerical example of the A-level count data, the first sample size $n = 9$ is very small, but the dotplot is not obviously non-normal. For the second sample, $n = 31$, so the assumption is less important. We observe some bunching, but there is also an indication of positive skewness. However, the coefficient of skewness (section 4.12) is only 0.45. As this is less than 1, we can conclude that skewness is not marked.

The second assumption requires $\sigma_1 = \sigma_2$ (i.e. $\sigma_1^2 = \sigma_2^2$). One way of seeing whether this is reasonable is to look at s_1 and s_2, the sample standard deviations. For our data, $s_1 = 7.10$ and $s_2 = 4.78$. This is not such good agreement, but is it good enough? The answer is 'yes' because an 'F test' indicates that the hypothesis that $\sigma_1^2 = \sigma_2^2$, should not be rejected. (Hypothesis tests are covered in Chapter 10 and the F test is covered in section 10.14.)

In order to carry out the calculations for a 95% confidence interval for $(\mu_1 - \mu_2)$, we note the following summary statistics:

$$\begin{aligned} \bar{x}_1 &= 16.22, \quad & s_1 &= 7.10, \quad & n_1 &= 9 \\ \bar{x}_2 &= 10.71, \quad & s_2 &= 4.78, \quad & n_2 &= 31. \end{aligned}$$

Then we obtain s using

$$\begin{aligned} s^2 &= \frac{(n_1 - 1)s_1^2 + (n_2 - 1)s_2^2}{n_1 + n_2 - 2} \\ &= \frac{(9 - 1)7.10^2 + (31 - 1)4.78^2}{9 + 31 - 2} \\ &= 28.65 \\ s &= \sqrt{28.65} = 5.35 \end{aligned}$$

So a 95% confidence interval for $(\mu_1 - \mu_2)$ is

$$(\bar{x}_1 - \bar{x}_2) \pm ts\sqrt{\frac{1}{n_1} + \frac{1}{n_2}}$$

$$(16.22 - 10.71) \pm 2.02 \times 5.35\sqrt{\frac{1}{9} + \frac{1}{31}}$$

$$5.51 \pm 4.09$$

$$1.4 \text{ to } 9.6$$

where $t = 2.02$ is taken from Table D.5 for $\alpha = 0.025$, $\nu = n_1 + n_2 - 2 = 9 + 31 - 2 = 38$. We are 95% confident that the difference between the population mean A-level counts of BA and BSc students is between 1.4 and 9.6. We note that the confidence interval does not contain the value zero, since both confidence limits have the same sign. Both limits in fact indicate that the mean A-level count for BA students is 'significantly higher' than for BSc students. We will return to this idea in the next chapter.

As in previous examples in this chapter, more precise estimates of $(\mu_1 - \mu_2)$ could have been obtained by taking larger samples.

Table 9.4 Confidence interval for $(\mu_1 - \mu_2)$, unpaired samples data

```
MTB> SET  C1
DATA> 32  22  . . .  10
DATA> END
MTB> SET  C2
DATA> 6  12 . . .
DATA> . . .
DATA> . . . 24
DATA> END
MTB> NAME  C1  'BACOUNT'
MTB> NAME  C2  'BSCOUNT'
MTB> TWOSAMPLE-T  95  C1  C2;
SUBC> POOLED.
```

Notes

(a) The data in column C1 and C2 are taken from column 7 of Appendix A.

(b) The command TWOSAMPLE-T also results in an 'unpaired *t* test' being carried out, the interpretation of which we will leave until section 10.13.

Minitab can do the above calculations (Table 9.4), but of course it will not tell you what assumptions to check for.

9.12 SUMMARY

A confidence interval for a parameter of a population, such as the mean, is a range within which we have a particular level of confidence, such as 95%, that the parameter lies. If we have randomly sampled data we can calculate confidence intervals for various parameters using one of the formulae in Table 9.5, but it is important to check whether the required assumptions are valid in each case.

We can also decide sample sizes if we can specify the precision with which we wish to estimate the parameter, and if we have some measure of variability from the results of a pilot experiment or survey.

WORKSHEET 9: CONFIDENCE INTERVAL ESTIMATION

1. Why are confidence intervals calculated?
2. What information do we need to calculate a 95% confidence interval

Table 9.5 Formulae for 95% confidence intervals

Parameter	*Case*	*Assumption*	*Formula*
μ Population mean	n large		$\bar{x} \pm 1.96\frac{s}{\sqrt{n}}$
μ Population mean	$n \geqslant 2$	Variable approximately normal	$\bar{x} \pm \frac{ts}{\sqrt{n}}$
p Binomial probability	$x > 5$ $n - x > 5$	The four conditions for a binomial distribution	$\frac{x}{n} \pm 1.96\sqrt{\frac{\frac{x}{n}\left(1 - n\right)}{n}}$
μ_d The mean of a population of differences	$n \geqslant 2$	Differences approximately normal	$\bar{d} \pm \frac{ts_d}{\sqrt{n}}$
$\mu_1 - \mu_2$ The difference in means of two populations	$n_1, n_2 \geqslant 2$	(i) Variable approximately normal	$(\bar{x}_1 - \bar{x}_2) \pm ts\sqrt{\frac{1}{n_1} + \frac{1}{n_2}}$ where
		(ii) $\sigma_1 = \sigma_2$	$s^2 = \frac{(n_1 - 1)s_1^2 + (n_2 - 1)s_2^2}{n_1 + n_2 - 2}$

for the mean of a population? What assumption is required if the sample size is small?

3. The larger the sample size, the wider the 95% confidence interval. True or false?
4. The more variation in the measurements, the wider the 95% confidence interval. True or false?
5. The higher the level of confidence, the wider the 95% confidence interval? True or false?
6. What does the following statement mean: 'I am 95% confident that the mean of the population lies between 10 and 12'.
7. Of a random sample of 100 customers who had not settled their accounts with an Electricity Board within one month of receiving them, the mean amount owed was £30 and the standard deviation

was £10. What is your estimate of the mean of all unsettled accounts? Suppose that the Electricity Board wanted an estimate of the mean of all unsettled accounts to be within £1 of the true figure for 95% confidence. How many customers who had not settled their accounts would need to be sampled?

8. Out of a random sample of 100 people, 80 said they were non-smokers. Estimate the percentage of non-smokers in the population with 95% confidence. How many people would need to be asked if the estimated percentage of non-smokers in the population is required to be within 1% for 95% confidence?

9. The systolic blood pressure of 90 normal British males has a mean of 128.9 mm of mercury and a standard deviation of 17 mm of mercury. Assuming these are a random sample of blood pressures, calculate a 95% confidence interval for the population mean blood pressure.
   (a) How wide is the interval?
   (b) How wide would the interval be if the confidence level were raised to 99%?
   (c) How wide would the 95% confidence interval be if the sample size were increased to 360?

   Are your answers to (a), (b) and (c) consistent with your answers to Questions 3 and 5 above?

10. In order to estimate the percentage of pebbles made of flint in a given locality to within 1% for 95% confidence, a pilot survey was carried out. Of a random sample of 30 pebbles, 12 were made of flint. How many pebbles need to be sampled in the main survey?

11. The number of drinks sold from a vending machine in a motorway service station was recorded on 60 consecutive days The results were as follows:

30	40	60	70	120	130	140	150	160	170
180	190	200	200	210	210	220	230	240	250
260	260	270	280	280	290	290	300	300	310
320	320	330	330	340	350	350	360	360	360
360	370	370	380	380	390	390	400	410	420
430	440	460	470	480	490	510	550	590	610

   Ignoring any differences between different days of the week and any time-trend or seasonal effects, estimate the mean number of drinks sold per day in the long term.

12. Ten women recorded their weights in kilograms before and after dieting. Assuming that the women were randomly selected, estimate the population mean reduction in weight. What additional assump-

tion is required and is it reasonable to make it here? The weights were:

Before	89.1	68.3	77.2	91.6	85.6	83.2	73.4	84.3	96.4	87.6
After	84.3	66.2	76.8	79.3	85.5	80.2	76.2	80.3	90.5	80.3

13. The percentage of a certain element in an alloy was determined for 16 specimens using one of two methods A and B. Eight of the specimens were randomly allocated to each method. The percentages were:

Method A	13.3	13.4	13.3	13.5	13.6	13.4	13.3	13.4
Method B	13.9	14.0	13.9	13.9	13.9	13.9	13.8	13.7

Calculate a 95% confidence interval for the difference in the mean percentages of the element in the alloy for the two methods, stating any assumptions made.

14. The annual rainfall in centimetres in two English towns over 11 years was as follows:

Year	*Town A*	*Town B*
1970	100	120
1971	89	115
1972	84	96
1973	120	115
1974	130	140
1975	105	120
1976	60	75
1977	70	90
1978	90	90
1979	108	105
1980	130	135

Estimate the mean difference in the annual rainfall for the two towns.

15. The actual weights of honey in 12 jars marked 452 g were recorded. Six of the jars were randomly selected from a large batch of brand A honey, and six were randomly selected from a large batch of brand B honey. The weights were:

Brand A	442	445	440	448	443	450
Brand B	452	450	456	456	460	449

Table 9.6 Minitab commands for 100 confidence intervals obtained from 100 samples of size 9 taken from a $N(70, 3^2)$ distributon

```
MTB> RANDOM  9  C1-C50;
SUBC> NORMAL  MU = 70  SIGMA = 3.
MTB> RANDOM  9  C51-C100;
SUBC> NORMAL  MU = 70  SIGMA = 3.
MTB> TINTERVAL  95  C1-C100
```

Estimate the mean difference in the weights of honey in jars marked 452 g for the two brands. Also estimate separately:
(a) The mean weight of brand A honey, and
(b) The mean weight of brand B honey.
Decide whether it is reasonable to suppose that the mean weight of honey from the brand A batch is 452 g, and similarly for brand B honey.

16. This question is designed to help you to understand more about confidence intervals using simulation on Minitab.

Confidence intervals are relatively simple to calculate, but what do they mean when we have calculated them? In the case of a 95% confidence interval for a population mean, μ, the answer is given by the last two sentences of section 9.2. (Similar statements can be made for confidence levels other than 95% and for other parameters such as μ_d and $\mu_1 - \mu_2$.) To illustrate the concept involved we can use simulation to take a number of samples of size n from a normal distribution with a known mean μ and standard deviation σ. For each sample we can then calculate a 95% confidence interval. If the formula used to calculate the 95% confidence intervals is correct, then we would expect that 95% of such intervals will 'capture' the mean μ. By 'capture' we mean that the known value of μ lies inside the confidence interval.

Table 9.6 is an example of such a simulation program. It will produce a total of 100 confidence intervals for μ, which we already know is 70. We can count how many intervals contain the value 70 and compare with the theory which indicates that 95 of the 100 intervals are expected to do so.

Hypothesis testing

> What tribunal can possibly decide truth in the clash of contradictory assertions and conjectures?

10.1 INTRODUCTION

Statistical inference is concerned with how we draw conclusions from sample data about the larger population from which the sample is selected. In the previous chapter we discussed one branch of inference, namely estimation, particularly confidence interval estimation. Another important branch of inference is hypothesis testing (Fig. 10.1), which is the subject of much of the remainder of this book.

In this chapter we consider all the cases we looked at in the previous chapter in terms of testing hypotheses about the various parameters, and conclude by discussing the connection between the two branches of inference.

The method of carrying out any hypothesis test can be set out in terms of seven steps:

1. Decide on a null hypothesis, H_0.
2. Decide on an alternative hypothesis, H_1.
3. Decide on a significance level.
4. Calculate the appropriate test statistic.
5. Find from tables the appropriate tabulated test statistic.
6. Compare the calculated and tabulated test statistics, and decide whether to reject the null hypothesis, H_0.
7. State a conclusion, and state the assumptions of the test.

In the following sections each of these steps and the concepts behind them will be explained, with the aid of an example.

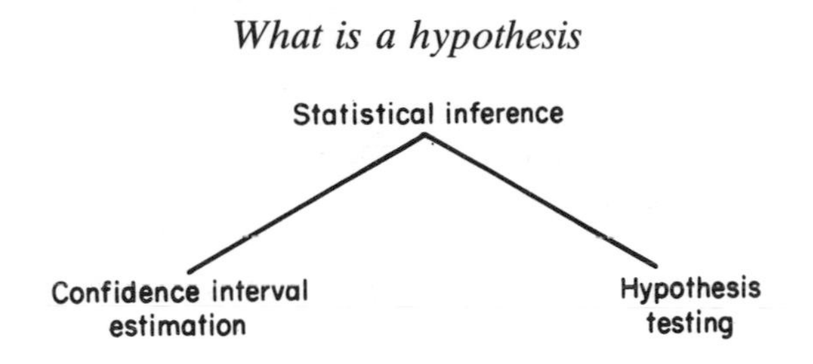

Fig. 10.1 Types of statistical inference.

10.2 WHAT IS A HYPOTHESIS?

In terms of the examples of the previous chapter, a hypothesis is a statement about the value of a population parameter, such as the population mean, μ. We use the sample data to decide whether the stated value of the parameter is reasonable. If we decide that it is not reasonable we reject the hypothesis in favour of another hypothesis. It is important to note at this stage, then, that in hypothesis testing we have two hypotheses to consider. Using sample data we decide which hypothesis is the more reasonable. We call the two hypotheses the **null hypothesis** and the **alternative hypothesis**.

10.3 WHICH IS THE NULL HYPOTHESIS AND WHICH IS THE ALTERNATIVE HYPOTHESIS?

The null hypothesis generally expresses the idea of 'no difference' – think of 'null' as meaning 'no'. In terms of the examples of the previous chapter a null hypothesis is a statement that the value of the population parameter is 'no different from', that is 'equal to', a specified value. The symbol we use to denote a null hypothesis is H_0. The null hypothesis

$$H_0\text{: } \mu = 165$$

states that the population mean equals 165.

The alternative hypothesis, which we denote by H_1, expresses the idea of 'some difference'. Alternative hypotheses may be **one-sided** or **two-sided**. The first two examples below are one-sided since each specifies only one side of the number 165; the third example is two-sided since both sides of 165 are specified:

$H_1\text{: } \mu > 165$ Population mean greater than 165
$H_1\text{: } \mu < 165$ Population mean less than 165
$H_1\text{: } \mu \neq 165$ Population mean not equal to 165

In a particular case we specify both the null and the alternative hypotheses according to the purpose of the investigation, and before the sample data are collected.

10.4 WHAT IS A SIGNIFICANCE LEVEL?

Hypothesis testing is also sometimes referred to as **significance testing**. The concept of **significance level** is similar to the concept of confidence level. The usual value we choose for our significance level is 5%, just as we usually choose a confidence level of 95%. Just as the confidence level expresses the idea that we would be prepared to bet heavily that the interval we state actually does contain the value of the population parameter of interest (see Worksheet 9, Question 6 and its solution), so a significance level of 5% expresses a similar idea in connection with hypothesis testing: a significance level of 5% is the risk we take in rejecting the null hypothesis, H_0, in favour of the alternative hypothesis H_1, when in reality H_0 is the correct hypothesis.

Example

If the first three steps of our hypothesis test (section 10.1) are:

1. H_0: $\mu = 165$
2. H_1: $\mu \neq 165$
3. 5% significance level

then we are stating that we are prepared to run a 5% risk that we will reject H_0 and conclude that the mean is not equal to 165, when the mean actually is equal to 165.

We cannot avoid the small risk of drawing such a wrong conclusion in hypothesis testing, because we are trying to draw conclusions about a population using only part of the information in the population, namely the sample data. The corresponding risk in confidence interval estimation is the small risk that the interval we state will not contain the true value of the population parameter of interest.

10.5 WHAT IS A TEST STATISTIC, AND HOW DO WE CALCULATE IT?

A test statistic is a value we can calculate from our sample data and the value we specify in the null hypothesis, using an appropriate formula.

Example

If the first three steps of our hypothesis test are as in the example of section 10.4, and our sample data are summarized as:

$$\bar{x} = 162.3, \quad s = 5.3, \quad n = 9$$

then the fourth step of our hypothesis test is as follows:

4. The calculated test statistic is:

$$Calc\ t = \frac{\bar{x} - \mu}{s/\sqrt{n}}$$

where μ refers to the value stated in the null hypothesis. (This formula for t was first introduced in section 9.5.)

$$= \frac{162.3 - 165}{5.3/\sqrt{9}}$$

$$= -1.53$$

10.6 HOW DO WE FIND THE TABULATED TEST STATISTIC?

We must know which tables to use for a particular application, and how to use them.

Example

1. H_0: $\mu = 165$.
2. H_1: $\mu \neq 165$.
3. 5% significance level.
4. *Calc* $t = -1.53$, from the previous section, assuming the same sample data ($\bar{x} = 162.3$, $s = 5.3$, $n = 9$).
5. The appropriate table is Table D.5, and we enter the tables for $\alpha = 0.05/2$, dividing by 2 since H_1 is two-sided. Here $\alpha = 0.05/2 = 0.025$, and $v = (n - 1) = 9 - 1 = 8$ degrees of freedom. So the tabulated test statistic is *Tab* $t = 2.306$ from Table D.5.

10.7 HOW DO WE COMPARE THE CALCULATED AND THE TABULATED TEST STATISTICS?

For the example in section 10.6 we reject H_0 if $|Calc\ t| > Tab\ t$, where the vertical lines mean that we ignore the sign of *Calc* t and consider only its magnitude (e.g. $|-5| = 5$, $|5| = 5$).

Since in this example $|Calc\ t| = 1.53$, and *Tab* $t = 2.306$, we do not reject H_0. Figure 10.2 shows that only calculated values of t in the tails of the distribution, beyond the *critical values* of -2.306 and 2.306, lead to the rejection of H_0.

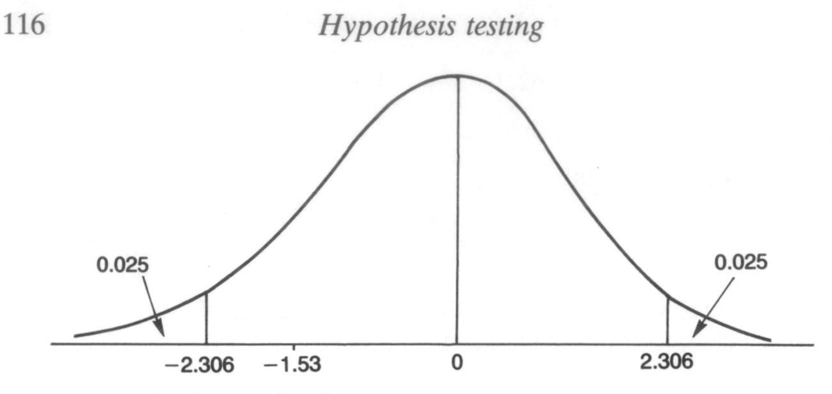

Fig. 10.2 t distribution for $\nu = 8$ degrees of freedom.

10.8 WHAT IS OUR CONCLUSION, AND WHAT ASSUMPTIONS HAVE WE MADE?

Our conclusion should be a sentence in words, as far as possible devoid of statistical terminology. For the example above, since we decided not to reject H_0 in favour of an alternative stating that the mean differed from 165, we conclude that 'the mean is not significantly different from 165 (5% level of significance)'.

The assumption of this test is that the variable is approximately normally distributed, which we have already seen (section 9.4) is reasonable for this example. Also, as we stated earlier, the assumption of normality is less critical the larger the sample size.

10.9 HYPOTHESIS TEST FOR THE MEAN, μ, OF A POPULATION

This section is a summary of the seven-step method in terms of the numerical example referred to in sections 10.3 to 10.8.

1. H_0: $\mu = 165$.
2. H_1: $\mu \neq 165$.
3. 5% significance level.
4. The calculated test statistic is:

$$Calc\ t = \frac{\bar{x} - \mu}{s/\sqrt{n}}$$

$$= \frac{162.3 - 165}{53/\sqrt{9}}$$

$$= -1.53$$

5. *Tab* $t = 2.306$, for $\alpha = 0.025$ and $\nu = 9 - 1 = 8$ degrees of freedom.
6. Since $|Calc\ t| < Tab\ t$, do not reject H_0.

7. The mean is not significantly different from 165 (5% level).
   Assumption: Variable is approximately normally distributed.

Notice that although we did not reject H_0, neither did we state that '$\mu = 165$'. We cannot be so definite, given that we have only sample data (μ refers to the population). The conclusion in Step 7 simply implies that we think H_0 is a more reasonable hypothesis than H_1 in this example. Clearly we cannot state '$\mu = 165$' and '$\mu = 164$', and so on.

10.10 TWO EXAMPLES OF HYPOTHESIS TESTS WITH ONE-SIDED ALTERNATIVE HYPOTHESES

If we had chosen a one-sided H_1 in the previous example, the steps would have varied a little. Since we could have chosen $\mu > 165$ or $\mu < 165$ as our alternative, both these examples are now given below, side by side:

1.	H_0: $\mu = 165$.	H_0: $\mu = 165$.
2.	H_1: $\mu > 165$.	H_1: $\mu < 165$.
3.	5% significance level.	5% significance level.
4.	*Calc* $t = -1.53$.	*Calc* $t = -1.53$.
5.	*Tab* $t = 1.860$ for $\alpha = 0.05/1$ $= 0.05$ and $\nu = (n - 1)$ $= 8$	*Tab* $t = 1.860$ for $\alpha = 0.05/1$ $= 0.05$ and $\nu = (n - 1)$ $= 8$
6.	Since *Calc t* < *Tab t*, do not reject H_0.	Since *Calc t* > − *Tab t*, do not reject H_0.
7.	The mean is not significantly greater than 165 (5% level).	The mean is not significantly less than 165 (5% level).

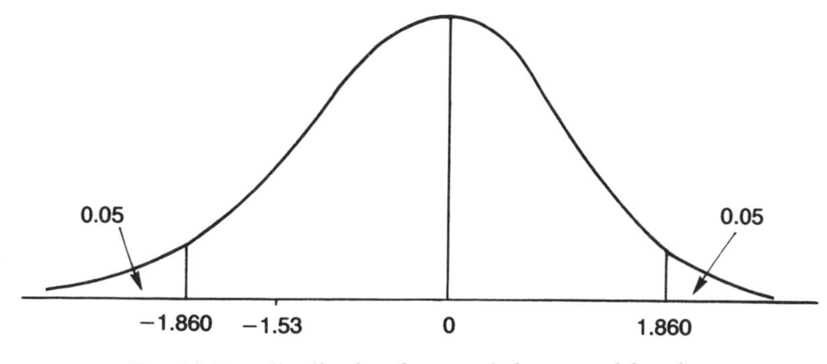

Fig. 10.3 t distribution for $\nu = 8$ degrees of freedom.

Assumption	Variable is approximately normally distributed. Figure 10.3 shows that only calculated values of t in the right-hand tail, greater than the critical value of 1.860, lead to rejection of H_0.	Variable is approximately normally distributed. Figure 10.3 shows that only calculated values of t in the left-hand tail, less than the critical value of − 1.860, lead to rejection of H_0.

Minitab commands and output for the 't test' in section 10.9 and the tests in section 10.10 are as follows:

Case 1. H_1: $\mu \neq 165$.

```
MTB>  TTEST  MU = 165 C1

     N  MEAN  STDEV     T      P
C1   9  162.3   5.3   -1.53  0.17
```

Case 2. H_1: $\mu > 165$.

```
MTB>  TTEST  MU = 165  C1;
SUBC> ALTERNATIVE = +1

N  MEAN  STDEV     T      P
9  162.3   5.3   -1.53  0.91
```

Case 3. H_1: $\mu < 165$.

```
MTB>  TTEST  MU = 165  C1;
SUBC> ALTERNATIVE = -1.

N  MEAN  STDEV     T      P
9  162.3   5.3   -1.53  0.086
```

We reject the null hypothesis, at the 5% level of significance, if the 'P value' is less than 0.05. In all three cases above, the 'P value' is greater than 0.05, so H_0 is not rejected in favour of H_1 in any of these cases.

Note that tests in which the alternative hypothesis is two-sided are often referred to as *two-tailed* tests, while tests in which the alternative hypothesis is one-sided are often referred to as *one-tailed* tests.

10.11 HYPOTHESIS TEST FOR A BINOMIAL PROBABILITY

Suppose we wish to test a hypothesis for p, the probability of success in a single trial, using sample data from a binomial experiment. If the experiment consisted of n trials, x of which were 'successes', the test statistic is calculated using the formula:

$$Calc\ z = \frac{\frac{x}{n} - p}{\sqrt{\frac{p(1-p)}{n}}}$$

where p is the value specified in the null hypothesis.

We can use this formula if $np > 5$, $n\ (1 - p) > 5$. The tabulated test statistic is *Tab z*, obtained from Table D.3(b).

Example

Test the hypothesis that the percentage of voters who will vote for party A in an election is 50%, against the alternative that it is greater than 50%, using the random sample data from an opinion poll that 110 out of 200 voters said they would vote for party A.

1. H_0: $p = 0.5$, which implies 50% vote for party A.
2. H_1: $p > 0.5$, which implies party A has an overall majority.
3. 5% significance level.
4. The test statistic

$$Calc\ z = \frac{\frac{x}{n} - p}{\sqrt{\frac{p(1-p)}{n}}}$$

can be used, since $np = 200 \times 0.5$ and $np(1 - p) = 200(1 - 0.5)$ are both greater than 5.

$$Calc\ z = \frac{\frac{110}{200} - 0.5}{\sqrt{\frac{0.5(1-0.5)}{200}}} = 1.414$$

5. *Tab z* = 1.645, since in Table D.3(b) this value of z corresponds to a tail of 0.05/1, the significance level divided by 1, because H_1 is one-sided (Fig. 10.4).
6. Since *Calc z* < *Tab z*, do not reject H_0.
7. The percentage of voters for party A is not significantly greater than 50% (5% level), so it is not reasonable to assume that party A will gain an overall majority in the election.
Assumption: The four binomial conditions apply (section 6.3).

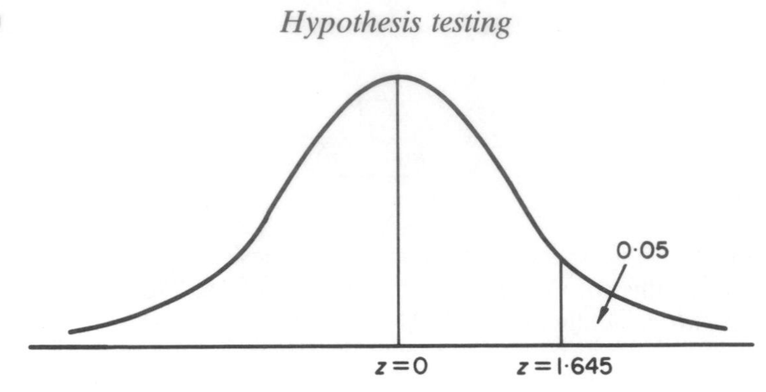

Fig. 10.4 Standardized normal distribution.

10.12 HYPOTHESIS TEST FOR THE MEAN OF A POPULATION OF DIFFERENCES, 'PAIRED' SAMPLES DATA

Example

For the example given in section 9.10 comparing two methods of teaching children to read, suppose we want to decide whether the new method, N, is better than the standard method, S, in terms of the mean difference in the test scores of the two methods.

We assume we have the same data as in Table 9.1, and hence the same summary of those data:

$$\bar{d} = 7.5, \quad s = 8.48, \quad n = 10$$

1. H_0: $\mu_d = 0$. This implies the mean of the population of differences is zero, that is, the methods give the same mean test score.
2. H_1: $\mu_d > 0$. Since the differences were calculated using N score − S score, this implies that the N method gives a higher mean test score.
3. 5% significance level.
4. The appropriate test statistic for this 'paired samples t test' is:

$$Calc\ t = \frac{\bar{d}}{s/\sqrt{n}}$$

$$= \frac{7.5}{8.48/\sqrt{10}}$$

$$= 2.80$$

5. $Tab\ t = 1.833$, for

$$\alpha = \frac{\text{sig. level}}{1}$$

since H_1 is one-sided

$$= 0.05$$

and $\nu = (n - 1) = 9$ degrees of freedom.

6. Since *Calc* $t >$ *Tab* t, reject H_0.
7. The N method gives a significantly higher mean test score than the S method (5% level).
   Assumption: The differences must be approximately normally distributed, and we saw in section 9.10 that this was a reasonable assumption to make for these data. Once again we can say that this assumption is less critical the larger the value of n.

Minitab commands and output for the test above, assuming that the data from Table 9.1 have been input and the differences calculated as in Table 9.2 and stored in C3, are as follows:

```
        MTB > TTEST  0  C3;
        SUBC> ALTERNATIVE + 1.
        N   MEAN   STDEV    T     P
    C3  10   7.5     8.48   2.80  0.010
```

Since the 'P value' is less than 0.05, we reject H_0 and conclude that the mean of the population of differences is greater than zero, i.e. that the N method gives a higher mean test score than the S method.

We note that the 95% confidence interval for μ_d was 1.4 to 13.6 (section 9.10), which also implies the same conclusion as above, since both confidence limits are positive, and recalling that the differences were calculated using difference = N score − S score.

10.13 HYPOTHESIS TEST FOR THE DIFFERENCE IN THE MEANS OF TWO POPULATIONS, 'UNPAIRED' SAMPLES DATA

Example

For the example given in section 9.11 comparing the A-level counts of BA and BSc students, suppose we want to test whether the mean A-level counts of the two populations are equal. We assume we have the same data as before which were summarized as follows:

$$\bar{x}_1 = 16.22, \quad s_1 = 7.10, \quad n_1 = 9$$
$$\bar{x}_2 = 10.72, \quad s_2 = 4.78, \quad n_2 = 31$$

The seven steps of the hypothesis test are as follows:

1. H_0: $\mu_1 = \mu_2$. There is no difference in the mean A-level counts of the two populations.

2. H_1: $\mu_1 \neq \mu_2$. There is a difference, in one direction or the other.
3. 5% significance level.
4. The appropriate test statistic for the 'unpaired samples t test' is

$$Calc\ t = \frac{\bar{x}_1 - \bar{x}_2}{s\sqrt{\frac{1}{n_1} + \frac{1}{n_2}}}$$

where, as in section 9.11,

$$s^2 = \frac{(n_1 - 1)s_1^2 + (n_2 - 1)s_2^2}{n_1 + n_2 - 2}$$

s^2 being the pooled estimate of variance. Using the data above, $s^2 = 28.65$, $s = 5.35$, and

$$Calc\ t = \frac{16.22 - 10.71}{5.35\sqrt{\frac{1}{9} + \frac{1}{31}}} = \frac{5.51}{2.03} = 2.72$$

5. *Tab* $t = 2.02$ for $\alpha = 0.05/2 = 0.025$, and $\nu = n_1 + n_2 - 2 = 9 + 31 - 2 = 38$ degrees of freedom.
6. Since $|Calc\ t| > Tab\ t$, i.e. $|2.72| = 2.72 > 2.02$, reject H_0.
7. The mean A-leval counts are significantly different (5% significance level). The direction of the difference is clear, since $\bar{x}_1 > \bar{x}_2$. Hence we can conclude that the mean A-level count for BA students is significantly higher than for BSc students. In fact we have already drawn this conclusion in section 9.11 by considering only the 95% confidence interval for $(\mu_1 - \mu_2)$. We discuss the connection between the topics of Chapters 9 and 10 in section 10.16 below.

   Assumptions: First, the measurements in each population must be approximately normally distributed, this assumption being less critical the larger the values of n_1 and n_2. Second, the population standard deviations, σ_1 and σ_2, must be equal. (For these data, both assumptions are reasonable, as discussed in section 9.11, since the same assumptions apply to the calculation of confidence intervals.)

The Minitab command TWOSAMPLE-T 95 C1 C2; and the subcommand POOLED. produce not only a 95% confidence interval for $(\mu_1 - \mu_2)$, but also the results of the hypothesis test as above, as follows:

```
MTB>  TWOSAMPLE-T 95   C1   C2;
SUBC> POOLED.

TWOSAMPLE T FOR C1 VS C2
      N   MEAN   STDEV   SEMEAN
C1    9   16.22   7.10     2.37
C2   31   10.71   4.78     0.86
```

95 PCT CI FOR MU C1 − MU C2: (1.4, 9.6)
POOLED SD = 5.35
TTEST MU C1 = MU C2 (VS NE): T = 2.72, P = 0.01, DF = 38

10.14 HYPOTHESIS TEST FOR THE EQUALITY OF THE VARIANCES OF TWO NORMALLY DISTRIBUTED POPULATIONS

In section 9.11 we obtained a confidence interval for $(\mu_1 - \mu_2)$, the difference between the means of two normally distributed populations. One of the assumptions needed to use the required formula was that the standard deviations of the two populations (σ_1 and σ_2) were equal. (This is the assumption we have just needed for the unpaired samples t test in section 10.13.) In section 9.11 we could not decide at the time whether this assumption was correct just by looking at the values of s_1 and s_2. The F test below provides a formal method of testing H_0: $\sigma_1^2 = \sigma_2^2$.

Example

For the sample data in section 9.11, we note that:

$$s_1 = 7.10, \quad n_1 = 9$$
$$s_2 = 4.78, \quad n_2 = 31$$

1. H_0: $\sigma_1^2 = \sigma_2^2$.
2. H_1: $\sigma_1^2 \neq \sigma_2^2$.
3. 5% significance level.
4. *Calc F* $= s_1^2/s_2^2$ if $s_1 > s_2$, or *Calc F* $= s_1^2/s_2^2$ if $s_1 > s_2$.
   For the above data, $s_1 > s_2$, so *Calc F* $= 7.10^2/4.78^2 = 2.21$.
5. *Tab F* $= 2.27$ using Table D.6 for $\nu_1 = 8$, $\nu_2 = 30$, where ν_1 is the number of d.f. associated with the numerator of *Calc F*, i.e. 7.10, $9 - 1 = 8$ d.f., and ν_2 is the number of d.f. associated with the denominator of *Calc F*, i.e. 4.78, $31 - 1 = 30$ d.f. We look up ν_1 along the top of the F table and ν_2 down the side.
6. Since *Calc F* < *Tab F*, H_0 is not rejected.
7. Hence the variances are not significantly different (5% level).
   Assumptions: Both populations are 'normal'. We were able to justify this using dotplots in section 9.11. However, if you require a less subjective test of normality than that afforded by dotplots, please refer to section 15.5.

10.15 THE EFFECT OF CHOOSING SIGNIFICANCE LEVELS OTHER THAN 5%

Why do we not choose a significance level lower than 5% since we would then run a smaller risk of rejecting H_0 when H_0 is correct (refer to section

10.4 if necessary)? Just as there are advantages and disadvantages in choosing a confidence level above 95% – a consequence of a higher confidence level is a wider confidence interval – a similar argument applies to significance levels below 5%.

If we reduce the significance level below 5% we reduce the risk of wrongly rejecting H_0, but we increase the risk of drawing a different wrong conclusion, namely the risk of wrongly rejecting H_1. Nor can we set both risks at 5% for the examples described in this chapter (for reasons which are beyond the scope of this book), and even if we could it might not be a wise thing to do! Consider the risks in a legal example and judge whether they should be equal:

(a) The risk of convicting an innocent man in a murder trial.
(b) The risk of releasing a guilty man in a murder trial.

There is nothing sacred about the '5%' for a significance level, nor the '95%' for a confidence level, but we should be aware of the consequences of departing from these conventional levels.

10.16 WHAT IF THE ASSUMPTIONS OF A HYPOTHESIS TEST ARE NOT VALID?

If at least one of the assumptions of a hypothesis test is not valid, i.e. there is insufficient evidence to make us believe they are all reasonable assumptions, then the test is also invalid and the conclusions may well be wrong.

In such cases, alternative tests, called **distribution-free** tests or more commonly **non-parametric** tests, should be used if they are available. These tests do not require such rigorous assumptions as the 'parametric' tests described in this chapter, but they have the disadvantage that they are less powerful, meaning that we are less likely to accept the alternative hypothesis when the alternative hypothesis is correct. Some non-parametric tests are described in Chapter 11.

10.17 THE CONNECTION BETWEEN CONFIDENCE INTERVAL ESTIMATION AND HYPOTHESIS TESTING

Confidence interval estimation and hypothesis testing provide similar types of information. However, a confidence interval (if a formula exists to calculate it) provides more information than the corresponding hypothesis test.

Example

Consider the student height data in the last example of section 9.4. Given sample data that $\bar{x} = 163.4$, $s = 5.3$, $n = 9$, we calculated a 95% confidence interval for the population mean, μ, of 158.0 to 166.2 cm.

From this result we can immediately state that any null hypothesis specifying a value of μ within this interval would not be rejected in favour of the two-sided alternative hypothesis, assuming a 5% level of significance. So, for example, H_0: $\mu = 160$ would not be rejected in favour of H_1: $\mu \neq 160$, but H_0: $\mu_0 = 150$ would be rejected in favour of H_1: $\mu \neq 150$, and H_0: $\mu = 170$ would be rejected in favour of H_1: $\mu \neq 170$.

We can state that 'a confidence interval for a parameter contains a range of values for the parameter we would not wish to reject'. The confidence interval is a way of representing all the null hypotheses we would not wish to reject, on the evidence of the sample data. To this extent a confidence interval contains much more information than the result of a hypothesis test.

10.18 SUMMARY

A statistical hypothesis is often a statement about the parameter of a population. In the seven-step method we use sample data to decide whether to reject the null hypothesis in favour of an alternative hypothesis. Table 10.1 summarizes the various tests covered in this chapter.

If the assumptions of a test are not valid, alternative non-parametric tests (to be discussed in Chapter 11) may be available.

The connection between confidence interval estimation and hypothesis testing was discussed: the former provides more information than the latter.

WORKSHEET 10: HYPOTHESIS TESTING INCLUDING t, z AND F TESTS

1. What is
   (a) A (statistical) hypothesis?
   (b) A null hypothesis?
   (c) An alternative hypothesis?
   Give an example of each in your main subject area.
2. What is a significance level?
3. Why do we need to run a risk of wrongly rejecting the null hypothesis?
4. Why do we choose 5% as the risk of wrongly rejecting the null hypothesis?

Table 10.1 Hypothesis tests

Parameter	*Case*	*Assumption*	*Decision rule for two-sided alternative hypothesis: Reject H_0 if*
μ Population mean	$n \geqslant 2$	Variable approximately normal	$\lvert Calc\ t\rvert = \dfrac{\lvert\bar{x} - \mu\rvert}{s/\sqrt{n}} > Tab\ t$
p Binomial probability	$np > 5$ $n(1 - p) > 5$	The four conditions for a binomial distribution	$\lvert Calc\ z\rvert = \dfrac{\left\lvert \dfrac{x}{n} - p\right\rvert}{\sqrt{\dfrac{p(1-p)}{n}}} > Tab\ z$
μ_d The mean of a population of differences	$n \geqslant 2$	Differences approximately normal	$\lvert Calc\ t\rvert = \dfrac{\lvert\bar{d}\rvert}{s_d/\sqrt{n}} > Tab\ t$
$\mu_1 - \mu_2$ The difference in means of two populations	$n_1, n_2 \geqslant 2$	(i) Variable is approximately normal (ii) $\sigma_1 = \sigma_2$	$\lvert Calc\ t\rvert = \dfrac{\lvert\bar{x}_1 - \bar{x}_2\rvert}{s\sqrt{\left(\dfrac{1}{n_1} + \dfrac{1}{n_2}\right)}} > Tab\ t$

5. How can we tell whether an alternative hypothesis is one-sided or two-sided?
6. How do we know whether to specify a one-sided or a two-sided alternative hypothesis in a particular investigation? Think of an example where each would be appropriate.

In Questions 7–18 inclusive, use a 5% significance level unless otherwise stated. In each question the assumptions required for the test should be stated, and you should also decide whether the assumptions are likely to be valid.

7. Eleven cartons of sugar, each nominally containing 1 kg, were randomly selected from a large batch of cartons. The weights of sugar were:

   1.02 1.05 1.08 1.03 1.00 1.06 1.08 1.01 1.04 1.07 1.00

   Do these data support the hypothesis that the mean weight for the batch is 1 kg?
8. A cigarette manufacturer claims that the mean nicotine content of a brand of cigarettes is 0.30 mg per cigarette. An independent con-

sumer group selected a random sample of 1000 cigarettes and found that the sample mean was 0.31 mg per cigarette, with a standard deviation of 0.03 mg. Is the manufacturer's claim justified or is the mean nicotine content significantly higher than he states?

9. The weekly take-home pay (£) of a random sample of 30 farm workers is as follows:

65	65	75	75	75	75	75	85	85	85
85	85	85	85	85	85	95	95	95	95
95	95	95	95	95	95	95	105	105	105

Do these data support the claim that the mean weekly take-home pay of farm workers is below £90?

10. The market share for the E10 size packet of a particular brand of washing powder has averaged 30% for a long period. Following a special advertising campaign it was discovered that, of a random sample of 100 people who had recently bought the E10 size packet, 35 had bought the brand in question. Has its market share increased?

11. Of a random sample of 300 gourmets, 180 prefer thin soup to thick soup. Is it reasonable to expect 50% of all gourmets prefer thin soup?

12. It is known from the records of an insurance company that 14% of all males holding a certain type of life insurance policy die during their 60th, 61st or 62nd year. The records also show that, of a randomly selected group of 1000 male civil servants holding this type of policy, 112 died during their 60th, 61st or 62nd year. Is it reasonable to suppose that male civil servants have a lower death rate than 14% during these years?

13. An experiment was conducted to compare the performance of two varieties of wheat, A and B. Seven farms were randomly chosen for the experiment, and the yields (in tonnes per hectare) for each variety on each farm were as follows:

Farm number	1	2	3	4	5	6	7
Yield of variety A	4.6	4.8	3.2	4.7	4.3	3.7	4.1
Yield of variety B	4.1	4.0	3.5	4.1	4.5	3.2	3.8

(a) Why do you think both varieties were tested on each farm, rather than testing variety A on seven farms and variety B on seven other farms?
(b) Carry out a hypothesis test to decide whether the mean yields are the same for the two varieties.

14. A sample length of material was cut from each of five randomly selected rolls of cloth and each length was divided into two halves. One half was dyed with a newly developed dye, and the other half with a dye that had been in use for some time. The ten pieces were then washed and the amount of dye washed out was recorded for each piece as follows:

Roll	1	2	3	4	5
Old dye	13.2	13.7	15.4	13.5	16.8
New dye	12.5	14.3	16.8	14.9	17.4

Investigate the allegation that the amount of dye washed out for the old dye is significantly less than for the new dye.

15. A sleeping drug and a neutral control were tested in turn on a random sample of ten patients in a hospital. The data below represent the differences between the number of hours of sleep under the drug and the neutral control for each patient:

$$2.0 \quad 0.2 \quad -0.4 \quad 0.3 \quad 0.7 \quad 1.2 \quad 0.6 \quad 1.8 \quad -0.2 \quad 1.0$$

Is it reasonable to assume that the drug would give more hours of sleep on average than the control for all the patients in the hospital?

16. To compare the strengths of two cements, six mortar cubes were made with each cement and the following strengths were recorded:

Cement A	4600	4710	4820	4670	4760	4480
Cement B	4400	4450	4700	4400	4170	4100

Is there a significant difference between the mean strengths of the two cements? In answering this question, state two assumptions you need to make, and test one of them. (Hint: use an F test first.)

17. The price of a standard 'basket' of household goods was recorded for 25 corner shops selected at random from the suburbs of all the cities in England. In addition, 25 food supermarkets were also randomly selected from the main shopping centres of the cities. The data were summarized as follows:

	Corner shop	*Supermarket*
Mean price (£)	9.45	8.27
Standard deviation of price (£)	0.60	0.50
Number in sample	25	25

Test the hypothesis that corner shops are charging the same on average as supermarkets for the standard basket.

18. Two geological sites were compared for the amount of vanadium (parts per million) found in clay. It was thought that the amount would be less for area A than for area B. Do the following data, which consist of ten samples taken randomly from each area, support this view?

Area A	*Area B*
50	50
65	70
75	90
80	95
90	95
95	100
105	105
110	110
130	125
140	150

19. Use the 95% confidence interval method on the data in Questions 7, 11, 13 and 16 to check the conclusions you reached using the hypothesis testing method.

Non-parametric hypothesis tests

At least the difference is very
inconsiderable.

11.1 INTRODUCTION

There are some hypothesis tests which do not require such rigorous assumptions as the tests described in Chapter 10. However, these 'non-parametric' tests are less powerful than the corresponding 'parametric' tests, that is we are less likely to reject the null hypothesis and hence accept the alternative hypothesis, when the alternative hypothesis is correct. We can therefore conclude that parametric tests are generally preferred if their assumptions are valid.

Three non-parametric tests are described in this chapter for which there are roughly equivalent t tests, as indicated in Table 11.1.

11.2 SIGN TEST FOR THE MEDIAN OF A POPULATION

Example

Suppose we collect the weekly incomes (£) of a random sample of ten self-employed window cleaners:

150 500 240 120 130 300 140 160 110 200

We are interested in testing a hypothesis concerning some average value of the data.

A dotplot of the data is shown in Fig. 11.1. It indicates positive skewness, and in fact using the measure given in section 4.12 we obstain a

Table 11.1 Non-parametric tests

Non-parametric test	*Application*	*Reference for roughly equivalenl t-test*
Sign test	Median of a population	Section 10.9
Sign test	Median of a population of differences – 'paired' samples data	Section 10.12
Wilcoxon signed rank test	Median of a population of differences – 'paired' samples data	Section 10.12
Mann–Whitney *U* test	Difference between the medians of two populations – 'unpaired' samples data	Section 10.13

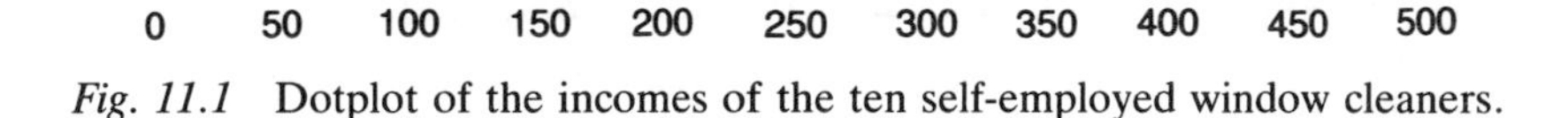

Fig. 11.1 Dotplot of the incomes of the ten self-employed window cleaners.

value of 1.26. Hence the assumption of normality is not reasonable, and a *t* test for the mean value would not be a valid test here. Let us therefore test the hypothesis that the median income is £200 per week. This can be done by carrying out a **sign test** as follows.

The sign test for the median of a population requires that we simply write down the signs of the differences between the incomes and the hypothesized value of £200. Using the convention that incomes above £200 have a + sign, and incomes below £200 have a − sign, and that incomes of exactly £200 (generally called *ties*) are disregarded, the signs for the above data are as follows:

$$- \quad + \quad + \quad - \quad - \quad + \quad - \quad - \quad -$$

There are three + signs, six − signs and one tie. If the null hypothesis that the median is £200 is true, the probability of an income above (or below) £200 is 1/2. We will take the alternative hypothesis to be that the median is not £200. The seven-step method for the sign test for this example is:

1. H_0: $p(+) = p(-) = 1/2$, where $p(+)$ and $p(-)$ mean the probability of a + sign and a − sign, respectively.
2. H_1: $p(+) \neq p(-)$, a two-sided alternative hypothesis.

3. 5% significance level.
4. For the sign test, the calculated test statistic is a binomial probability. In general, it is the probability of getting the result obtained *or a result which is more extreme*, assuming, for the purposes of calculating the probability, that H_0 is true. For the example we need to calculate:

$$\begin{aligned}
&\mathrm{P}(6 \textit{ or more} \text{ minus signs in 9 trials for } p(-) = 1/2)\\
&= \mathrm{P}(6) + \mathrm{P}(7) + \mathrm{P}(8) + \mathrm{P}(9)\\
&= \binom{9}{6}\left(\frac{1}{2}\right)^6\left(\frac{1}{2}\right)^3 + \binom{9}{7}\left(\frac{1}{2}\right)^7\left(\frac{1}{2}\right)^2 + \binom{9}{8}\left(\frac{1}{2}\right)^8\left(\frac{1}{2}\right)^1 + \binom{9}{9}\left(\frac{1}{2}\right)^9\left(\frac{1}{2}\right)^0\\
&= \left(\frac{9 \times 8 \times 7}{1 \times 2 \times 3} + \frac{9 \times 8}{1 \times 2} + \frac{9}{1} + 1\right)\left(\frac{1}{2}\right)^9\\
&= (84 + 36 + 9 + 1)\left(\frac{1}{2}\right)^9\\
&= 0.2539.
\end{aligned}$$

5. The tabulated test statistic for the sign test is simply the significance level divided by 2, if H_1 is two-sided, and equals 0.025 for this example.
6. Reject H_0 if the calculated probability is less than (significance level)/2, for a two-sided H_1. For this example, since $0.2539 > 0.025$, we do not reject H_0.
7. The median wage is not significantly different from £200 (5% level). *Assumption*: The variable has a continuous distribution.

Notes

(a) Instead of P(6 or more minus signs in 9 trials for $p(-) = 1/2$), we could have calculated P(3 or fewer plus signs in 9 trials for $p(+) = 1/2$), but the answer would have been the same, because of the symmetry of the binomial when $p = 1/2$.

(b) Notice that the assumption of a continuous distribution is much less restrictive than the assumption of a normal distribution.

(c) If $n > 10$, we can alternatively use the method of section 11.4.

(d) Since we did not use the magnitudes of the differences (between each income and £200) this test can be performed even if we do not have the actual differences, but simply their signs.

11.3 SIGN TEST FOR THE MEDIAN OF A POPULATION OF DIFFERENCES, 'PAIRED' SAMPLES DATA

Example

For the example given in section 9.10 of two methods of teaching children to read, suppose we want to decide whether the new method (N) is better than the standard method (S), but we do not wish to assume that the differences in the test scores are normally distributed. Instead we can use the sign test to decide whether the median score by the new method is significantly greater than that by the standard method.

The differences (N score − S score) were:

$$7 \quad -2 \quad 6 \quad 4 \quad 22 \quad 15 \quad -5 \quad 1 \quad 12 \quad 15.$$

1. H_0: $p(+) = p(-) = 1/2$. The median of the population of differences is zero, i.e. median of N scores equals the median of S scores.
2. H_1: $p(+) > p(-)$. Median of N scores is greater than the median of S scores.
3. 5% significance level.

4 and 5. If the null hypothesis is true we would expect equal numbers of + and − signs. If the alternative hypothesis is true we would expect more + signs, so the null hypothesis is rejected if:

P(observed number *or more* of + signs out of 10 for $p(+) = 1/2$) < 0.05, for a one-sided H_1.

In the example we have 8 plus signs in 10 differences.

$$\begin{aligned} \text{P}(8 \textit{ or more}) &= 1 - \text{P}(7 \text{ or fewer}) \\ &= 1 - 0.9543 \\ &= 0.0547, \text{ using Table D.1.} \end{aligned}$$

6. Since $0.0547 > 0.05$ (if only just), we do not reject the null hypothesis, H_0.
7. The median score by the N method is not significantly greater than that by the S method (5% level).
   Assumption: The differences are continuous, which can be assumed for the test scores even if we quote them to the nearest whole number.

Notes

(a) If $n > 10$, we can alternatively use the method of section 11.4.

(b) This test can be used in cases where we do not know or cannot quantify the magnitudes of the differences, for example in preference testing. Only the signs of the differences are used, so we could use the convention: brand A preferred to brand B is recorded as a +, and so on.

(c) Differences of zero are ignored in this test.

11.4 SIGN TEST FOR LARGE SAMPLES ($n > 10$)

The sign test for sample sizes larger than 10 is made easier by the use of a normal approximation method (similar to that used in section 7.7) by putting:

$$\mu = \frac{n}{2} \quad \text{and} \quad \sigma = \frac{\sqrt{n}}{2}$$

Example

Suppose that for $n = 30$ paired samples there are 20 + and 10 − differences.

1. H_0: $p(+) = p(-) = 1/2$. The median of the population of differences is zero.
2. H_1: $p(+) \neq p(-)$ (two-sided).
3. 5% significance level.
4. Following the method used in the example in section 11.2, we now need to calculate P(20 or more + signs in 30 trials, for $p(+) = 1/2$). Instead of calculating several binomial probabilities we can put

$$\mu = \frac{n}{2} = 15, \quad \sigma = \frac{\sqrt{n}}{2} = 2.74$$

and use the corresponding normal distribution (Fig. 11.2).

We also need a continuity correction since '20 or more on a discrete scale' is equivalent to 'more than 19.5 on a continuous scale' (remembering that the binomial is for discrete variables, and the normal distribution is for continuous variables).

For $x = 19.5$, *Calc* $z = (x - \mu)/\sigma = (19.5 - 15)/2.74 = 1.64$.

5. *Tab* $z = 1.96$ from Table D.3(b), since this value of z corresponds to a tail area of 0.05/2, the significance level divided by 2 since H_1 is two-sided.
6. Since $|Calc\ z| < Tab\ z$, we do not reject the null hypothesis.
7. The median of differences is not significantly different from zero (5% level).

Assumption: The differences are continuous.

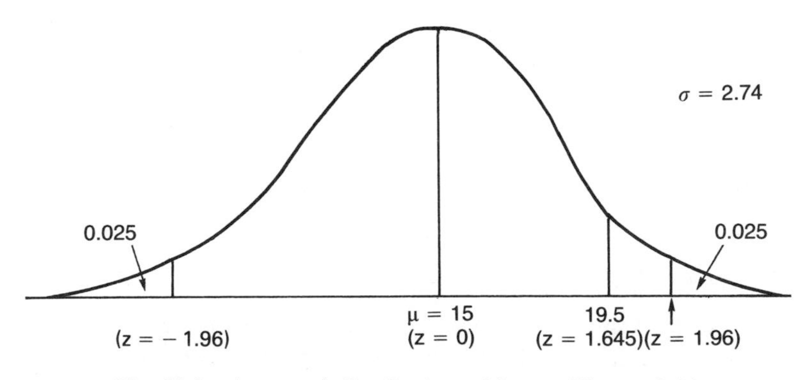

Fig. 11.2 A normal distribution with $\mu = 15$, $\sigma = 2.74$.

11.5 SIGN TEST USING MINITAB

The Minitab commands and output for the sign test examples shown above in sections 11.2 and 11.3 are given in Tables 11.2 and 11.3, respectively. In fact Minitab uses the large sample test of section 11.4, so the conclusions could have differed from those given earlier, but here they are in agreement.

For the window cleaners data in Table 11.2, since the '*P* value' is greater than 0.05, the null hypothesis that the median is 200 is not rejected, as we showed in section 11.2.

For the teaching methods data in Table 11.3, since the '*P* value' is again greater than 0.05, H_0 is not rejected, as in section 11.3.

11.6 WILCOXON SIGNED RANK TEST FOR THE MEDIAN OF A POPULATION OF DIFFERENCES, 'PAIRED' SAMPLES DATA

In the Wilcoxon signed rank test the hypothesis tested is the same as for the sign test. Since the former test uses the magnitudes as well as the

Table 11.2 A sign test for the median of a population

```
MTB> SET  C1
DATA> 150  500  240  120  130  300  140  160  110  200
DATA> END
MTB> NAME  C1  'INCOMES'
MTB> STEST  200  C1

SIGN TEST OF MEDIAN = 200 VERSUS MEDIAN N.E. 200
              N    BELOW    EQUAL    ABOVE      P     MEDIAN
INCOMES      10        6        1        3   0.51        155
```

Table 11.3 A sign test for the median of a population of differences

```
MTB> SET  C1
DATA> 7  -2  6  4  22  15  -5  1  12  15
DATA> END
MTB> NAME  C1  'DIFFS'
MTB> STEST  0  C1;
SUBC> ALTERNATIVE + 1.

SIGN TEST OF MEDIAN = 0.00 VERSUS G.T. 0.00
            N    BELOW    EQUAL    ABOVE       P     MEDIAN
DIFFS      10        2        0        8   0.055        6.5
```

signs of the differences, it is more powerful than the sign test, and hence is a preferred method when the magnitudes are known.

The general method of obtaining the calculated test statistic for the Wilcoxon signed rank test is as follows. Disregarding ties (a tie is a difference of zero), the n differences are ranked without regard to sign. The sum of the ranks of the positive differences, T_+, and the sum of the ranks of the negative differences, T_-, are calculated. The smaller of T_+ and T_- is the calculated test statistic, T. (A useful check is $T_+ + T_- = \frac{1}{2}n(n + 1)$.)

Example

Using the data of section 11.3, the differences are as follows:

Differences (N score − S score)	7	−2	6	4	22	15	−5	1	12	15
Ordering the differences without regard to sign	1	−2	4	−5	6	7	12	15	15	22
The corresponding ranks are	1	2	3	4	5	6	7	$8\frac{1}{2}$	$8\frac{1}{2}$	10

Observe the instance of tied ranks. The two values in rank positions 8 and 9 are equal (to 15) and are both given the mean of the ranks they would have been given if they had differed slightly.

We now obtain a value of T:

$$\begin{aligned} T_+ &= \text{sum of the ranks of the + differences} \\ &= 1 + 3 + 5 + 6 + 7 + 8\tfrac{1}{2} + 8\tfrac{1}{2} + 10 = 49 \\ T_- &= \text{sum of the ranks of the − differences} \\ &= 2 + 4 = 6 \end{aligned}$$

Adding these sums together, we have $T_+ + T_- = 49 + 6 = 55$, and $\frac{1}{2}n(n + 1) = 1/2 \times 10 \times 11 = 55$, so this agrees. The smaller of T_+ and T_- is 6, so $T = 6$.

Setting out the seven-step method:

1. H_0: The median of the population of differences is zero, which implies the median of N scores is equal to the median of the S scores.
2. H_1: The median of N scores is greater than the median of S scores.
3. 5% significance level.
4. *Calc* $T = 6$, from above.
5. *Tab* $T = 10$, from Table D.7 of Appendix D for 5% significance level, one-sided H_1, and $n = 10$.
6. Since *Calc* $T <$ *Tab* T is true here, reject H_0.
7. The median of N scores is significantly greater than the median of S scores (5% level).

Assumption: The distribution of the differences is continuous and symmetrical.

Notes

(a) In step 6 we reject H_0 if *Calc* $T \leqslant$ *Tab* T, i.e. even if *Calc* $T =$ *Tab* T.

(b) When $n > 25$, Table D.7 cannot be used. Instead we use the method of section 11.7 in this case.

(c) The same data have been analysed using both the sign test and the Wilcoxon signed rank test. However, the conclusions are not the same! Using the sign test H_0 was not rejected (although the decision was a close one) and using the Wilcoxon test H_0 was rejected. Since, as we have already mentioned, the latter test is more powerful, the latter conclusion is preferred.

11.7 WILCOXON SIGNED RANK TEST FOR LARGE SAMPLES ($n > 25$)

When $n > 25$, Table D.6 cannot be used. Instead we use a normal approximation method by putting:

$$\mu_T = \frac{n(n+1)}{4} \quad \text{and} \quad \sigma_T = \sqrt{\frac{n(n+1)(2n+1)}{24}}$$

Example

Suppose that for $n = 30$ paired samples $T_+ = 300$, $T_- = 165$, so that $T = 165$.

1. H_0: The median of the population of differences is zero.
2. H_1: The median of the population of differences is not zero (two-sided).
3. 5% significance level.
4. To obtain the test statistic we first need values of μ_T and σ_T:

$$\mu_T = \frac{n(n+1)}{4} = 232.5, \quad \sigma_T = \sqrt{\frac{n(n+1)(2n+1)}{24}} = 48.6,$$

The normal distribution with these parameters is shown in Fig. 11.3. We can now find *Calc* z:

$$Calc\ z = \frac{T - \mu_T}{\sigma_T} = \frac{165.5 - 232.5}{48.6} = 1.38$$

Note the use of the continuity correction as in section 11.4.

5. *Tab* $z = 1.96$ from Table D.3(b), since this value of z corresponds to a tail area of 0.05/2, the significance level divided by 2 because H_1 is two-sided.

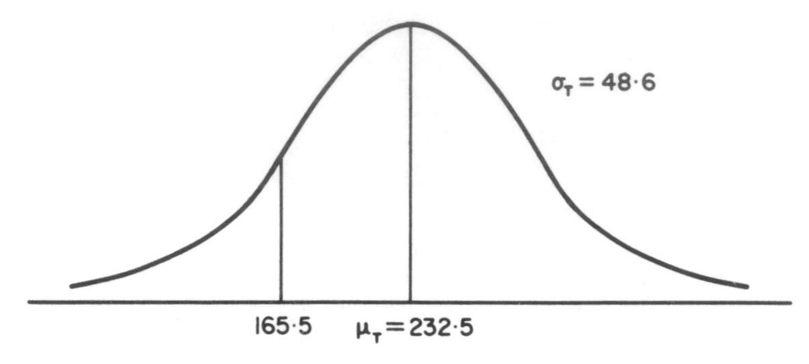

Fig. 11.3 A normal distribution with $\mu_T = 232.5$, $\sigma_T = 48.6$.

6. Since $|Calc\ z| < Tab\ z$, we do not reject H_0.
7. The median of differences is not significantly different from zero (5% level).

   Assumption: The distribution of differences is continuous and symmetrical.

11.8 WILCOXON SIGNED RANK TEST USING MINITAB

The Minitab commands and output for the Wilcoxon signed rank test example shown in section 11.6 are given in Table 11.4. Once again, Minitab uses the large sample test of section 11.7, so the conclusions could have differed from those given earlier, but in fact they are in agreement. Since the '*P* value' is less than 0.05, H_0 is rejected, as in section 11.6.

Table 11.4 A Wilcoxon signed rank test for the median of a population of differences

```
MTB> SET  C1
DATA> 7  −2  6  4  22  15  −5  1  12  15
DATA> END
MTB> NAME  C1  'DIFFS'
MTB> WTEST  0  C1;
SUBC> ALTERNATIVE + 1.

WILCOXON TEST OF MEDIAN = 0.00 versus G.T. 0.00
            N        TEST STATISTIC         P          MEDIAN
CI          10            49.0            0.016          7.0
```

11.9 MANN–WHITNEY U TEST FOR THE DIFFERENCE IN THE MEDIANS OF TWO POPULATIONS, 'UNPAIRED' SAMPLES DATA

If we cannot justify the assumptions required in the unpaired samples t test (section 10.13), the Mann–Whitney U test may be used to test the following hypotheses:

H_0: The two populations have distributions which are identical in all respects.
H_1: The two populations have distributions with different medians, but are otherwise identical.

The alternative hypothesis here is two-sided, but one-sided alternatives can also be specified.

The general method of obtaining the calculated test statistic for the Mann–Whitney U test is as follows. Letting n_1 and n_2 be the sizes of the samples drawn from the two populations, the $(n_1 + n_2)$ sample observations are ranked as one group. The sums of the ranks of the observations in each sample, R_1 and R_2, are calculated. Then U_1 and U_2 are calculated using:

$$U_1 = n_1 n_2 + \tfrac{1}{2} n_1 (n_1 + 1) - R_1$$
$$U_2 = n_1 n_2 + \tfrac{1}{2} n_2 (n_2 + 1) - R_2.$$

(A useful check is $U_1 + U_2 = n_1 n_2$.) The smaller of U_1 and U_2 is the calculated test statistic, U.

Example

As part of an investigation into factors underlying the capacity for exercise, a random sample of 11 factory workers took part in an exercise test. Their heart rates in beats per minute at a given level of oxygen consumption were as follows:

112 104 109 107 149 127 125 152 103 111 132

A random sample of nine racing cyclists also took part in the same exercise test and their heart rates were:

91 111 115 123 83 112 115 84 120

These data are plotted in Fig. 11.4, which is similar to a dotplot with the dots replaced by the actual values to facilitate ranking. If we plotted the data on two accurate dotplots, they would not look convincingly normal. A Mann–Whitney U test is called for.

Ranking all 20 observations as one group, giving equal values the average of the ranks they would have had if they had differed slightly, we obtain Fig. 11.5. We now calculate U as follows:

Factory workers, ($n_1 = 11$)		103,104,107,109,111,112		125,127 132 149,152
Cyclists ($n_2 = 9$)	83,84 91		111,112,115 120 123 115	

Fig. 11.4 Heart rates of factory workers and cyclists.

Factory workers ($n = 11$)		4,5,6,7,$8\frac{1}{2}$,$10\frac{1}{2}$		16,17 18 19,20
Cyclists ($n = 9$)	1,2, 3		$8\frac{1}{2}$,$10\frac{1}{2}$,$12\frac{1}{2}$ 14,15 $12\frac{1}{2}$	

Fig. 11.5 Ranks of heart rates of factory workers and cyclists.

$$n_1 = 11, \quad n_2 = 9,$$
$$R_1 = 4 + 5 + 6 + 7 + 8\tfrac{1}{2} + 10\tfrac{1}{2} + 16 + 17 + 18 + 19 + 20 = 131$$
$$R_2 = 1 + 2 + 3 + 8\tfrac{1}{2} + 10\tfrac{1}{2} + 12\tfrac{1}{2} + 12\tfrac{1}{2} + 14 + 15 = 79$$
$$U_1 = n_1n_2 + \tfrac{1}{2}n_1(n_1 + 1) - R_1 = 11 \times 9 + \tfrac{1}{2} \times 11 \times 12 - 131 = 34$$
$$U_2 = n_1n_2 + \tfrac{1}{2}n_2(n_2 + 1) - R_2 = 11 \times 9 + \tfrac{1}{2} \times 9 \times 10 - 79 = 65.$$
$$\text{Check:} \quad U_1 + U_2 = 34 + 65 = 99, \quad n_1n_2 = 99.$$

The smaller of U_1 and U_2 is 34, so $U = 34$.
Setting out the seven-step method,

1. H_0: The populations of the heart rates for factory workers and cyclists have identical distributions.
2. H_1: The distributions have different medians, but are otherwise identical (two-sided).
3. 5% significance level.
4. *Calc U* = 34, from above.
5. *Tab U* = 23, from Table D.8 of Appendix D for a 5% significance level, two-sided H_1, $n_1 = 11$, $n_2 = 9$.
6. Since *Calc U* > *Tab U*, do not reject H_0.
7. The median heart rates for factory workers and cyclists are not significantly different (5% level).
   Assumption: The variable is continuous. Since the number of beats per minute is large and may be the average of several observations, this assumption is reasonable.

Notes

(a) In step 6 we reject H_0 if *Calc U* $\leqslant$ *Tab U*, i.e. even if *Calc U* = *Tab U*.

(b) When n_1 or n_2 is greater than 20, Table D.8 cannot be used. Instead we use the method of section 11.10.

11.10 MANN–WHITNEY *U* TEST FOR LARGE SAMPLES

When n_1 or n_2 is greater than 20, we use a normal approximation method by putting:

$$\mu_U = \frac{n_1 n_2}{2} \quad \text{and} \quad \sigma_U = \sqrt{\frac{n_1 n_2(n_1 + n_2 + 1)}{12}}$$

Example

Suppose that for two unpaired samples of size $n_1 = 25$, $n_2 = 30$, we obtain $R_1 = 575$, $R_2 = 965$, $U_1 = 500$, $U_2 = 250$, so $U = 250$.

1. H_0: The two populations have identical distributions.
2. H_1: The two populations have distributions with different medians, but are otherwise identical (two-sided).
3. 5% significance level.
4. We calculate μ_U and σ_U as above:

$$\mu_U = \frac{n_1 n_2}{2} = 375, \quad \sigma_U = \sqrt{\frac{n_1 n_2(n_1 + n_2 + 1)}{12}} = 59.2$$

   The normal distribution with these parameters is shown in Fig. 11.6. We can now calculate the required test statistic:

$$\textit{Calc } z = \frac{U - \mu_U}{\sigma_U} = \frac{250.5 - 375}{59.2} = -2.10.$$

   Note the use of the continuity correction as in section 11.4.
5. *Tab* $z = 1.96$ from Table D.3(b), since this value of z corresponds to a tail area of 0.05/2, the significance level divided by 2, since H_1 is two-sided.

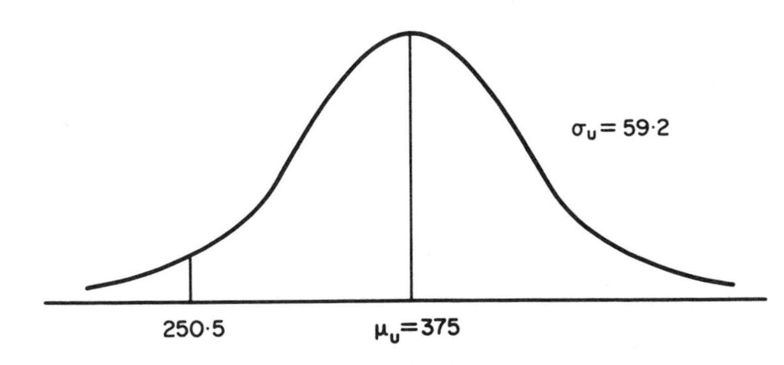

Fig. 11.6 A normal distribution with $\mu_U = 375$, $\sigma_U = 59.2$.

6. Since $|Calc\ z| > Tab\ z$, reject H_0.
7. The medians are significantly different (5% level).
   Assumption: The variable is continuous.

11.11 MANN–WHITNEY *U* TEST USING MINITAB

The Minitab commands and output for the Mann–Whitney *U* test example shown in section 11.9 are given in Table 11.5. Once more, Minitab uses the large sample test of section 11.8, so the conclusions could have differed from those given earlier, but in fact they are the same. Since the '*P* value' is greater than 0.05, H_0 is not rejected, as in section 11.9.

11.12 SUMMARY

Three non-parametric tests, namely the sign test, the Wilcoxon signed rank test and the Mann–Whitney *U* test are described for the small- and large-sample cases. These tests make less rigorous assumptions than the corresponding *t* tests, but are less powerful if the assumptions of the *t* tests are valid. Table 11.6 summarizes the various tests covered in this chapter.

WORKSHEET 11: SIGN TEST, WILCOXON SIGNED RANK TEST, MANN–WHITNEY *U* TEST

Fill in the gaps in Questions 1, 2, 3 and 5.

1. Non-parametric tests are used to test _____ in cases where the _____ of the corresponding parametric tests are not valid.

Table 11.5 A Mann–Whitney *U* test for the difference between two medians

```
MTB> SET  C1
DATA> 112  104  109  107  149  127  125  152  103  111  132
DATA> END
MTB> SET  C2
DATA> 91  111  115  123  83  112  115  84  120
DATA> END
MTB> NAME C1  'WORKERS'
MTB> NAME C2  'CYCLISTS'
MTB> MANNWHITNEY  C1  C2

TEST OF ETA1 = ETA2 VS. ETA1 N.E. ETA2 IS SIGNIFICANT AT 0.2545.
THE TEST IS SIGNIFICANT AT 0.2539 (ADJUSTED FOR TIES).
CANNOT REJECT AT ALPHA = 0.05.
```

Table 11.6 Non-parametric hypothesis tests

Test	*Parameter*	*Case*	*Assumption*	*Reject* H_0 *if**
Sign	Median of a population	$n \leqslant 10$	Variable has a continuous distribution	*Calc binomial probability* $\leqslant \frac{\text{significance level}}{2}$
Sign	Median of a population of differences	$n \leqslant 10$	Variable has a continuous distribution	*Calc binomial probability* $\leqslant \frac{\text{significance level}}{2}$
Sign	Median of a population of differences	$n > 10$	Variable has a continuous distribution	$\lvert Calc\ z \rvert > Tab\ z$
Wilcoxon	Median of a population of differences	$n \leqslant 25$	Variable has a continuous and symmetrical distribution	$Calc\ T \leqslant Tab\ T$
Wilcoxon	Median of a population of differences	$n > 25$	Variable has a continuous and symmetrical distribution	$\lvert Calc\ z \rvert > Tab\ z$
Mann–Whitney	Difference between the medians of two populations	$n_1, n_2 \leqslant 20$	Variable has a continuous distribution	$Calc\ U \leqslant Tab\ U$
Mann–Whitney	Difference between the medians of two populations	$n_1, n_2 > 20$	Variable has a continuous distribution	$\lvert Calc\ z \rvert > Tab\ z$

* The 'decision rules' in this column are for two-sided alternative hypotheses only, except for T and U.

2. However, when the _____ are valid it is better to use parametric tests because they are more _____ than the corresponding non-parametric tests.

3. Power is the risk of rejecting the _____ hypothesis when the _____ hypothesis is correct. The _____ the power of a hypothesis test, the better.

4. The sign test and the Wilcoxon signed rank test may both be used on paired samples data. Give examples of data which could
   (a) only be analysed using the sign test,
   (b) be analysed using either test. Which test is preferable in this case?

5. The Mann–Whitney U test is a non-parametric test which corresponds to the _____ _____ t test. The latter is a more _____ test if two _____ are valid. These are that:
   (a) both variables are _____ distributed;
   (b) the _____ _____ of the two populations are equal.

6. Reanalyse the data from Worksheet 10, Question 7, using the sign test, and compare the conclusion with that of the t test.

7. What further information would you need for the data in Worksheet 10, Question 8, in order to carry out a sign test?

8. Reanalyse the data from Worksheet 10, Question 14, using the Wilcoxon signed rank test.

9. A psychologist tested eight subjects, randomly chosen from the 11-year-old boys taught in the comprehensive schools in a city, using a standard aptitude test. The scores were:

   135 103 129 96 122 140 110 91

   (a) Later the same subjects received a new (improved!) aptitude test and the scores (in the same order of subject) were:

   125 102 117 94 120 130 110 92

   Is there a significant difference between the average scores for the two tests? Use an appropriate non-parametric test.
   (b) Now assume that the scores in the second test refer to an independent second random sample of eight subjects. Is there a significant difference between the average scores for the two tests? Again use an appropriate non-parametric test.

10. An investigation was carried out on a trout farm to find the effect of a new feeding compound. Twenty fry (newly born fish) were randomly divided into two equal groups. Both groups were then kept under

the same environmental conditions, but one group was fed with a standard feeding compound and the other group was fed with the new feeding compound. After a given period the fish were weighed. Their weights (in grams) were as follows:

Standard compound	*New compound*
510	521
507	476
490	489
496	512
523	521
508	498
534	505
497	547
491	542
506	492

Analyse these data using a non-parametric test.

11. Two brands of car tyre were tested in simulated road trials. The 'distances' travelled by 12 tyres of one brand and 12 tyres of the other brand before their treads had worn below the legal minimum limit were recorded to the nearest thousand kilometres:

Brand 1	47	44	39	41	39	42	51	44	55	45	49	46
Brand 2	43	33	40	38	31	39	34	40	35	37	38	32

Is one brand better than the other? Use a non-parametric test.

Association of categorical variables

12.1 INTRODUCTION

The inferential methods discussed in Chapters 9, 10 and 11 involved data for one variable measured on a number of 'individuals', where the variable was numerical (either continuous, discrete, or ranked). We now turn to data for categorical variables (refer to section 1.2). Also, instead of one variable (univariate) data, we will discuss two-variable (bivariate) data.

So, in this chapter, we will be concerned with investigations in which two categorical variables are recorded for a number of 'individuals'. Such data may be set out in two-way **contingency tables** (refer to section 3.4). We will decide whether the two variables are independent or whether they are associated, by performing a hypothesis test, namely the χ^2 (chi-squared) test for independence.

12.2 CONTINGENCY TABLES

Remember that a categorical variable is one which is not quantitative, but can take 'values' which are non-numerical categories or classes. So, for an individual, 'type of subject' and 'reaction to the flavour of a new toothpaste' are two examples of categorical variables. The categories for the first variable could be 'adult' and 'child', while for the second variable they could be 'liked flavour' and 'disliked flavour'. We can represent the numbers of individuals falling into each cross-category in a contingency table.

Consider the results of a market research survey to find the reaction of adults and children to the flavour of a new toothpaste (Table 12.1). Since this table has two rows and two columns, it is called a 2×2 contingency table. One with r rows and c columns is called an $r \times c$ contingency table. If we wish to know whether there is significant association between the

Table 12.1 Contingency table for reaction to toothpaste and type of subject

	Type of subject	
Reaction	*Adult*	*Child*
Liked flavour	90	100
Disliked flavour	50	30

two variables, or whether the variables are independent, we can carry out a χ^2 test of independence, given certain conditions, as described in the next section.

12.3 χ^2 TEST FOR INDEPENDENCE, 2 × 2 CONTINGENCY TABLE DATA

We will carry out the usual seven-step method of hypothesis testing:

1. H_0: The variables 'type of subject' and 'reaction to toothpaste' are independent, i.e. there is no association between the variables.
2. H_1: The variables are not independent, they are associated (two-sided).
3. 5% significance level.
4. If the observed frequencies are denoted O, and if all expected frequencies, E, are greater than or equal to 5 (see note (a) below), then the calculated test statistic is

$$Calc\ \chi^2 = \sum \frac{(O - E)^2}{E}$$

for the general $r \times c$ table, but we use

$$Calc\ \chi^2 = \sum \frac{(|O - E| - \frac{1}{2})^2}{E}$$

for the 2 × 2 table. The latter formula includes what is called **Yates's continuity correction**. The values in Table 12.1 are examples of observed frequencies. The expected frequencies are those we would expect if we assume (for the purpose of calculating them) that the null hypothesis is correct and we keep the row and column totals (as in Table 12.2) fixed. We calculate the E values for each cell in the contingency table using the formula:

$$E = \frac{\text{row total} \times \text{column total}}{\text{grand total}}$$

Table 12.2 Observed (and expected) frequencies

	Type of subject		
Reaction	*Adult*	*Child*	*Row total*
Liked flavour	90 (98.5)	100 (91.5)	190
Disliked flavour	50 (41.5)	30 (38.5)	80
Column total	140	130	Grand total 270

It is convenient to put the E values in brackets next to the corresponding O values (Table 12.2). For example, we calculate the E value for row 2, column 1 as follows:

$$E = \frac{\text{row 2 total} \times \text{col 1 total}}{\text{grand total}} = \frac{80 \times 140}{270} = 41.5$$

The symbol Σ in the formula for *Calc* χ^2 means that we sum over all cells in the table. The term $|O - E|$ in the formula for a 2×2 table means that we take the magnitude of the $(O - E)$ values and ignore the sign, for example $|90 - 98.5| = |-8.5| = 8.5$, and $|90 - 98.5| - \frac{1}{2} = 8.5 - \frac{1}{2} = 8.0$.

For the data in Table 12.2,

$$\begin{aligned} Calc\ \chi^2 &= \frac{(|90 - 98.5| - \frac{1}{2})^2}{98.5} + \frac{(|100 - 91.5| - \frac{1}{2})^2}{91.5} \\ &\quad + \frac{(|50 - 41.5| - \frac{1}{2})^2}{41.5} + \frac{(|30 - 38.5| - \frac{1}{2})^2}{38.5} \\ &= \frac{(8.5 - 0.5)^2}{98.5} + \frac{(8.5 - 0.5)^2}{91.5} + \frac{(8.5 - 0.5)^2}{41.5} + \frac{(8.5 - 0.5)^2}{38.5} \\ &= 4.55. \end{aligned}$$

5. *Tab* χ^2 is obtained from Table D.9, and we enter the tables for α = significance level, even though H_1 is two-sided, and $\nu = (r - 1)(c - 1)$. For the example of a 5% significance level and a 2×2 table, $\alpha = 0.05$, $\nu = (2 - 1)(2 - 1) = 1$ and *Tab* $\chi^2 = 3.84$.
6. If *Calc* $\chi^2 >$ *Tab* χ^2, reject the null hypothesis. For the example, *Calc* $\chi^2 = 4.55$, *Tab* $\chi^2 = 3.84$, so the null hypothesis is rejected.
7. We conclude that there is significant association between type of subject and reaction to the toothpaste (5% level). The 'direction' of the association is clear if we look at the individual cells of Table 12.2. For example, fewer adults than expected – under the null hypothesis

of independence – like the flavour, while the opposite is true for children. We can state that 'a significantly higher proportion of children like the flavour compared with adults (5% level)'.

Notes

The following notes relate to the above example and to the analysis of contingency table data in general. Please read them carefully!

(a) All the expected values must be at least 5, as stated in step 4 of the above test. If this is not the case we may:
  (i) perform a different test if we have a 2×2 table (section 12.6), or
  (ii) collapse rows (if $r > 2$) or columns (if $c > 2$) to form a smaller table, but only if it makes sense to do so (section 12.4).

(b) The observations in the contingency table must be independent. The best way to ensure this is by taking a random sample. An example of dependent data might be if each child in the toothpaste example was related to one of the adults, thus forming pairs of subjects.

(c) The observations in the contingency table must be frequencies, not percentages, proportions or measurements.

(d) The null hypothesis of independence may also be expressed in terms of proportions (or probabilities). So for the data in Table 12.2 we could write:

$$H_0\text{: P(liked flavour} \mid \text{adult)} = \text{P(liked flavour} \mid \text{child)}$$

(refer to section 5.9 on conditional probabilities, if necessary).

(e) The fact that we may conclude that there is significant association between the variables does not necessarily imply cause and effect (we will make a similar statement in connection with significant correlation coefficients in Chapter 13). So we cannot conclude that being a child causes an individual to like the flavour, nor can we conclude that liking the flavour causes an individual to become a child.

(f) The formula for the degrees of freedom for the χ^2 test can be justified in terms of the definition of section 9.7. For a 2×2 table, once we have calculated one E value, the other three are determined by the restriction that the row and column totals are fixed (the sum of the E values in any row, or column, is the same as the sum of the O values). So for a 2×2 table there is only one degree of freedom, and a similar argument can be applied to justify the use of $(r - 1)(c - 1)$ degrees of freedom for the general $r \times c$ table.

12.4 χ^2 TEST FOR INDEPENDENCE, 3×3 TABLE

Because so many points were discussed in the previous example, particularly notes (a) and (b), we now consider another example, this time for a 3×3 table.

Example

Sixty workers were randomly selected and asked to give their opinion on a new pension scheme which their employers were considering. Of 10 workers with a high income, 8 were in favour of the new scheme, 1 was undecided and 1 was against the new scheme. Of 25 workers with an average income, the numbers in favour, undecided and against were 7, 3 and 15, respectively. Of 25 workers with a low income, the corresponding numbers were 2, 10 and 13, respectively. Are the opinions of workers independent of their income? (Although income is quantitative it is treated as categorical for the purposes of this example.)

Table 12.3 can be formed from the above information. E values (calculated as in section 12.3) are given in brackets to one decimal place, which is sufficiently accurate.

Since we have three E values less than 5 in this table we cannot apply the χ^2 test to these data as they stand. We can, however, combine the top two rows, as in Table 12.4. It makes sense to combine rows here because income is a ranked variable and also the top row is where the low E values are. However, it would not make sense to combine rows 1 and 3!

Table 12.3 3 × 3 contingency table for income and opinion of 60 workers

	Opinion on a new pension scheme			
Income	*In favour*	*Undecided*	*Against*	*Row total*
High	8 (2.8)	1 (2.3)	1 (4.8)	10
Average	7 (7.1)	3 (5.8)	15 (12.1)	25
Low	2 (7.1)	10 (5.8)	13 (12.1)	25
Column total	17	14	29	60

Table 12.4 2 × 3 contingency table for income and opinion of 60 workers

	Opinion on a new pension scheme			
Income	*In favour*	*Undecided*	*Against*	*Row total*
High or Average	15 (9.9)	4 (8.2)	16 (16.9)	35
Low	2 (7.1)	10 (5.8)	13 (12.1)	25
Column total	17	14	29	60

With all E values now greater than 5, the seven-step method can now be applied:

1. H_0: Opinion (on a new pension scheme) and income are independent.
2. H_1: Opinion and income are associated (two-sided).
3. 5% significance level.
4. The calculated test statistic for the 2 × 3 table is:

$$Calc\ \chi^2 = \sum \frac{(O - E)^2}{E}$$

$$= \frac{(15 - 9.9)^2}{9.9} + \frac{(4 - 8.2)^2}{8.2} + \frac{(16 - 16.9)^2}{16.9} + \frac{(2 - 7.1)^2}{7.1} + \frac{(10 - 5.8)^2}{5.8} + \frac{(13 - 12.1)^2}{12.1}$$

$$= 11.6$$

5. $Tab\ \chi^2 = 5.99$ for $\alpha = 0.05$, $\nu = (2 - 1)(3 - 1) = 2$.
6. Since $Calc\ \chi^2 > Tab\ \chi^2$, we reject the null hypothesis.
7. Opinion and income are significantly associated. Looking at the cells with the greatest contribution to the $Calc\ \chi^2$, we conclude that:
   (a) more of the high or average income are in favour, and
   (b) more of the low income workers are undecided,
   than we would expect if income and opinion were independent.

Note

The categories of the variable 'Opinion on a new pension scheme', namely in favour, undecided and against, are in a logical ranking order. In this situation, a further test, called a χ^2 trend test, is appropriate (section 12.7).

12.5 χ^2 TEST FOR INDEPENDENCE USING MINITAB

The Minitab commands for carrying out a χ^2 test for independence using the data in Table 12.1 are shown in Table 12.5.

Table 12.5 χ^2 test for the example in section 12.2

```
MTB> READ  C1  C2
DATA> 90  100
DATA> 50  30
DATA> END
MTB> CHISQUARE  C1  C2
```

Notes

(a) The output to the χ^2 test gives only the value of *Calc* χ^2, i.e. no '*P* value' is given.

(b) Minitab does not use Yates's correction in the case of a 2×2 table, which is why Minitab's value of *Calc* χ^2 is 5.14 for this example, compared with 4.55 as stated in section 12.2. Although both values are greater than 3.84 (so the null hypothesis is rejected in each case), my advice is to use Yates's correction for any 2×2 table.

12.6 FISHER EXACT TEST

As stated in note (a) after the example in section 12.2, if not all the E values are at least 5, the χ^2 test is invalid. If this is the case for a 2×2 table we may use a different test, called the **Fisher exact test**. The method is as follows.

Suppose that we have the following 2×2 table of observed frequencies, one of which is less than 5:

a	b	$a + b$
c	d	$c + d$
$a + c$	$b + d$	n

where we have also included marginal (row and column) totals, and n is the sum of the observed frequencies, $a + b + c + d$.

We first calculate the probability

$$\frac{(a + b)!\,(c + d)!\,(a + c)!\,(b + d)!}{n!\,a!\,b!\,c!\,d!}$$

Assuming a two-sided alternative hypothesis, this procedure is repeated for all 2×2 tables with the same marginal totals. All the probabilities thus obtained which are less than or equal to the initial probability, including the initial probability itself, are then summed to obtain the total probability. The null hypothesis is rejected if this total probability is less than 0.05, assuming a 5% significance level.

Example

Forty students (see Appendix A) were classified according to their sex and the type of degree for which they were studying (see Table 12.6, which is the same as Table 3.9). We will test the null hypothesis that sex and type of degree are independent, against a two-sided alternative. Under H_0, the expected values are as follows, using the formula given in section 12.3:

Table 12.6 Contingency table for sex and type of degree for forty students

	Type of degree	
Sex	*BA*	*BSc*
Male	2	11
Female	7	20

2.9	10.1
6.1	20.9

Since 2.9 < 5, the χ^2 test is invalid, and a Fisher exact test is called for.

For the data in Table 12.6, $a = 2$, $b = 11$, $c = 7$, $d = 20$, $n = 40$, so

$$\text{probability} = \frac{13!\,27!\,9!\,31!}{40!\,2!\,11!\,7!\,20!} = 0.2533$$

Since 0.2533 is already greater than 0.05, the null hypothesis is not rejected, and there is no need to carry out any further probability calculations, which can only make the total probability larger. However, for illustration purposes only, the other tables with the same marginal totals and their corresponding probabilities (in parentheses) are:

1 12	0 13	3 10	4 9	5 8	6 7	7 6	8 5	9 4
8 19	9 18	6 21	5 22	4 23	3 24	2 25	1 26	0 27
(0.1055)	(0.0171)	(0.3096)	(0.2111)	(0.0826)	(0.0184)	(0.0022)	(0.0001)	(0.0000)

(the total probability is, of course, 1, the total of all those less than or equal to 0.2533 is about 0.69 which is, of course, still greater than 0.05).

The formal steps in the Fisher exact test are as follows:

1. H_0: sex and type of degree are independent.
2. H_1: sex and type of degree are not independent (two-sided).
3. 5% significance level.
4. *Calc probability* = 0.69, or we could just say 'greater than 0.2533' based on the initial table only.
5. 0.05 is the 'critical' probability (there is no 'tabulated' probability).
6. Since 0.69 or 0.2533 > 0.05, H_0 is not rejected.
7. Sex and type of degree are independent.

Notes

(a) As with the χ^2 test of independence, the observations must be independent.

(b) For a one-sided H_1, we consider only those tables which are more extreme in the direction of the alternative hypothesis. For example, if the alternative hypothesis had been 'males are more likely to study for a BSc', we would consider only the tables with observed frequencies of 2, 1, and 0 in the top left cell of the table. This gives a total probability of 0.2533 + 0.1055 + 0.0171 to be compared with 0.05 as above.

(c) The Fisher exact test is not available on Minitab.

12.7 χ^2 TREND TEST

This test was mentioned in the note at the end of section 12.4. It should be used in conjunction with the standard χ^2 test when one of the variables has more than two ordered categories. This was the case in Table 12.4 where the variable 'opinion on a new pension scheme' had three ordered categories. These are scored −1, 0 and +1, respectively, as shown in the following example, in which you should be able to follow the method even though formulae are not given!

The rationale behind the trend test is that the standard χ^2 test for independence takes no account of the ordered categories, if they are present.

Example

Perform a χ^2 trend test on the data in Table 12.4.

The observed frequencies are:

	15	4	16	35
	2	10	13	25
	17	14	29	60
and we write down scores	−1	0	1	

We now calculate:

$$15 \times (-1) + 4 \times 0 + 16 \times 1 = 1$$
$$17 \times (-1) + 14 \times 0 + 29 \times 1 = 12$$
$$17 \times (-1)^2 + 14 \times 0^2 + 29 \times 1^2 = 46$$

$$Calc\ \chi_1^2 = \frac{60\{60 \times 1 - 35 \times 12\}^2}{35 \times 25\{60 \times 46 - 12^2\}} = \frac{7\,776\,000}{2\,289\,000} = 3.4$$

We now calculate $Calc\ \chi_2^2 = 11.6$, using the standard χ^2 test (section 12.4). Then we calculate

$$Calc\ \chi_3^2 = Calc\ \chi_2^2 - Calc\ \chi_1^2 = 11.6 - 3.4 = 8.2$$

This is significant at the 5% level, since 8.2 > 3.84. The conclusion is that only a small non-significant part of the total χ^2 is explained by a linear trend. The majority of the total χ^2 is explained by non-linear differences between observed and expected frequencies. We can see this more clearly by noting the proportions of high or average incomes for each of the three types of opinion, i.e. 88%, 29% and 55%. There is no hint of a linear trend either upwards or downwards in these figures.

12.8 SUMMARY

Inferences from bivariate categorical data were discussed. Such data, collected from a random sample of 'individuals', may be displayed as observed frequencies (O) in two-way contingency tables. The null hypothesis of independence between the variables is tested by means of the χ^2 statistic where:

$$Calc\ \chi^2 = \sum \frac{(O - E)^2}{E}$$

if the number of degrees of freedom, $(r - 1)(c - 1)$, are greater than 1, or

$$Calc\ \chi^2 = \sum \frac{(|O - E| - \frac{1}{2})^2}{E}$$

if the number of degrees of freedom equals 1. The expected frequencies (E) are calculated using:

$$E = \frac{\text{row total} \times \text{column total}}{\text{grand total}}$$

If any E value is less than 5, the formula for *Calc* χ^2 is invalid, and alternative methods must be considered. These include collapsing tables (if $r > 2$ or $c > 2$) or using the Fisher exact test in the case of 2×2 tables. Rejection of the null hypothesis of independence does not imply cause and effect. If the categorical variable has more than two categories and they are logically ordered a χ^2 trend test should be performed.

WORKSHEET 12: ASSOCIATION OF CATEGORICAL VARIABLES

Fill in the gaps in Questions 1–7.

1. A categorical variable can only take 'values' which are non-______.
2. If we collect data for two categorical variables for a number of individuals, the data may be displayed in a two-way or ______ table. In such a table the numbers in the various cells of the table are the

number of _____ in each cross-category and are referred to as _____ frequencies.

3. The null hypothesis in the analysis of contingency table data is that the two categorical variables are _____.

4. In order to calculate the χ^2 statistic we first calculate the _____ frequencies, using the formula:

$$E = \_\_\_\_\_\_\_\_.$$

5. If all the E values are greater than or equal to _____, the test statistic *Calc* χ^2 = _____ is calculated. Since the E values are calculated assuming the null hypothesis is true, high values of *Calc* χ^2 will tend to lead to the _____ of the null hypothesis.

6. The degrees of freedom for *Tab* χ^2 are (_____)(_____) for a contingency table with r rows and c columns, so for a 2×2 contingency table there are _____ degrees of freedom.

7. For a 2×2 contingency table we reject the null hypothesis at the 5% level of significance if *Calc* $\chi^2 >$ _____.

8. Of 40 rented television sets, the tubes of 9 sets burnt out within the guarantee period of two years. Of 60 bought sets, the tubes of 5 sets burnt out within two years. Test the hypothesis that the proportion of burnt sets is independent of whether they were bought or rented, assuming that the 100 referred to are a random sample of all sets.

9. For four garages in a city selling the same brand of four-star petrol the following are the numbers of male and female car drivers calling for petrol between 5 p.m. and 6 p.m. on a given day. Is there any evidence that the proportion of male to female varies from one garage to another?

Sex of driver	*Garages*				
	A	*B*	*C*	*D*	*Totals*
Male	25	50	20	25	120
Female	10	50	5	15	80
Totals	35	100	25	40	200

10. The examination results of 50 students, and their attendance (%) on a course, were as follows:

	Exam result		
Attendance	*Pass*	*Fail*	*Totals*
Over 70%	20	5	25
30%–70%	10	5	15
Under 30%	5	5	10
Totals	35	15	50

Is good attendance associated with a higher chance of passing the examination?

11. Two types of sandstone were investigated for the presence of three types of mollusc. The numbers of occurrences were:

	Type of mollusc		
	A	*B*	*C*
Sandstone 1	15	30	12
Sandstone 2	15	9	6

Is there enough evidence to suggest that the proportion of the three types of mollusc is different for the two types of sandstone?

12. In a survey of pig farms it is suspected that the occurrence of a particular disease may be associated with the method of feeding. Methods of feeding are grouped into two categories, A and B. Of 5 farms on which the disease occurred, 4 used method A and 1 method B. Of 15 farms on which the disease had not occurred, 9 used method A and 6 method B. Test for independence between the method of feeding and the occurrence of the disease.

13. Two drugs, denoted by A and B, were tesed for their effectiveness in treating a certain common mild illness. Of 1000 patients suffering from the illness, 700 were chosen at random and given drug A, and the remaining 300 were given drug B. After one week, 100 of the patients were worse, 400 showed no change in their condition, and 500 were better. On the assumption that the two drugs are identical in their effect, complete a table similar in form to that below to show, for each drug, the expected number of patients getting worse, showing no change, and becoming better.

The given table shows the observed number of patients in each category. Carry out a χ^2 test, at the 5% level, to determine whether the six observed frequencies are consistent with the assumption of identical effects.

Also carry out a χ^2 trend test, and state your overall conclusions.

	Number of patients		
	Becoming worse	*Showing no change*	*Becoming better*
Drug A	64	255	381
Drug B	36	145	119

Correlation of quantitative variables

> Besides, in many instances it is impossible to determine whether these are causes or effects.

13.1 INTRODUCTION

In the previous chapter we discussed tests of the association of two categorical variables. If, instead, we are interested in the association of two quantitative (numerical) variables, we may:

(a) summarize the sample data graphically in a scatter diagram (refer to Fig. 3.10);
(b) calculate a numerical measure of the degree of association, called a correlation coefficient;
(a) carry out a hypothesis test that there is no correlation in the bivariate population from which the sample data were drawn, and interpret the conclusion of this test with great care!

We will discuss two correlation coefficients: Pearson's (product moment) correlation coefficient (sections 13.2 and 13.3), which we use if we can reasonably assume that each of the two variables is separately normally distributed; and Spearman's rank correlation coefficient (sections 13.4–13.7), which we use if we cannot reasonably assume normality. If the variables are normally distributed the hypothesis test for Pearson's coefficient is more powerful than the test for Spearman's coefficient.

13.2 PEARSON'S CORRELATION COEFFICIENT

Suppose we record the heights and weights of a random sample of six adults (Table 13.1). It is reasonable to assume that these variables are

Table 13.1 Heights (cm) and weights (kg) of a random sample of six adults

Height	*Weight*
170	57
175	64
176	70
178	76
183	71
185	82

normally distributed, based on past experience of such variables, so the Pearson correlation coefficient is the appropriate measure of the degree of association between height and weight.

A scatter diagram for these data is shown in Fig. 13.1.

We will discuss this scatter diagram later once we have calculated the value of Pearson's correlation coefficient – we will use the symbol r to represent the value of this coefficient obtained from sample data, and ρ (rho) to represent the value for the population. The formula for r is:

$$r = \frac{\Sigma xy - \dfrac{\Sigma x \Sigma y}{n}}{\sqrt{\left[\Sigma x^2 - \dfrac{(\Sigma x)^2}{n}\right]\left[\Sigma y^2 - \dfrac{(\Sigma y)^2}{n}\right]}}$$

where one of our variables is the x variable, the other is the y variable, and n is the number of individuals. In correlation it is an arbitrary decision as to which variable we call x and which we call y. Suppose we decide that weight is the x variable and height is the y variable as in Fig. 13.1; then in the formula for r, Σx means the sum of the weights, and so on.

For the data in Table 13.1,

$$\begin{aligned}
\Sigma x &= 57 + 64 + 70 + 76 + 71 + 82 &&= 420\\
\Sigma x^2 &= 57^2 + 64^2 + \cdots + 82^2 &&= 29\,786\\
\Sigma y &= 170 + 175 + \cdots + 185 &&= 1\,067\\
\Sigma y^2 &= 170^2 + 175^2 + \cdots + 185^2 &&= 189\,899\\
\Sigma xy &= (57 \times 170) + (64 \times 175) + \cdots + (82 \times 185) &&= 74\,901\\
n &= 6
\end{aligned}$$

$$r = \frac{74\,901 - \dfrac{420 \times 1067}{6}}{\sqrt{\left(29\,786 - \dfrac{420^2}{6}\right)\left(189\,899 - \dfrac{1067^2}{6}\right)}}$$

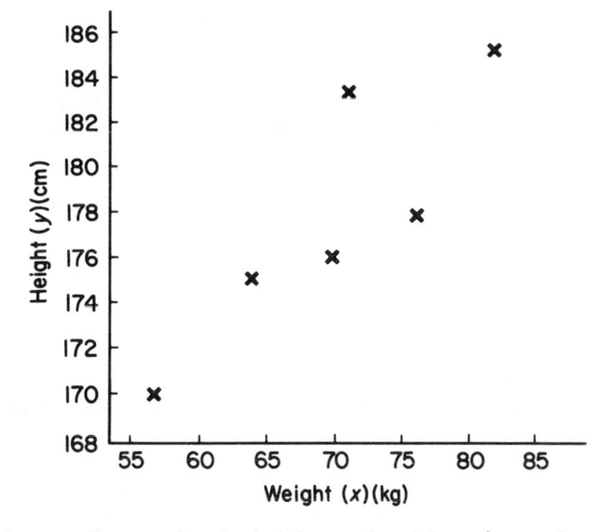

Fig. 13.1 Scatter diagram for the heights and weights of a random sample of six adults.

$$= \frac{211}{\sqrt{386 \times 150.8}}$$
$= 0.874$. (Check through this calculation if it is unfamiliar.)

What does a sample correlation coefficient of 0.874 tell us? In order to put this value in perspective we can look at the scatter diagram of Fig. 13.1, where the general impression is of increasing weight to be associated with increasing height, and vice versa. We can imagine a cigar-shaped outline round the data to emphasize this impression.

In general, if points on a scatter diagram show the same tendency (i.e. as one variable increases, so does the other) and in addition the points lie on a straight line, then $r = 1$. If there is the opposite tendency (i.e. as one variable increases, the other decreases) and in addition the points lie on a straight line, then $r = -1$. If there is no such tendency and the points look like the distribution of cherries in a perfectly made cherry cake, then the value of r is close to 0. These three cases are shown in Fig. 13.2.

Within the range of possible values for r from -1 to $+1$, we may describe a value of 0.874, as obtained above, as 'high positive correlation'. But be warned. Do not judge the association between two variables simply from the value of the correlation coefficient! We must also take into account n, the number of 'individuals' contributing to the sample data. Intuitively, $r = 0.874$ based on a sample of only $n = 6$ individuals is not as impressive as $r = 0.874$ based on a sample of $n = 60$ individuals. Had we obtained the latter we would have more evidence of the degree

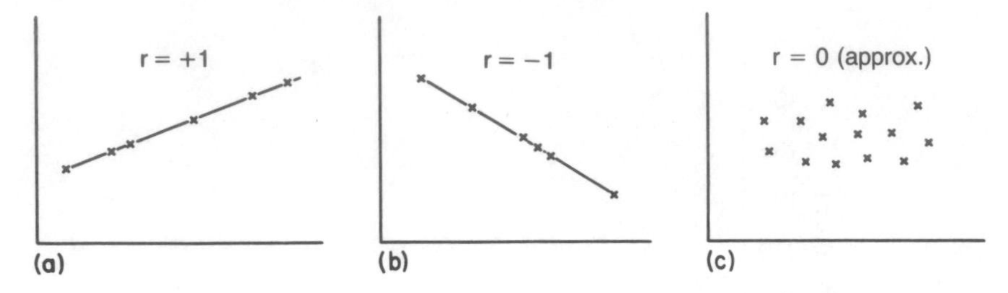

Fig. 13.2 Scatter diagrams for: (a) $r = +1$; (b) $r = -1$; (c) $r = 0$ (approx.).

of association in the population. This intuitive argument is formalized in a hypothesis test for the population coefficient in the next section.

13.3 HYPOTHESIS TEST FOR PEARSON'S POPULATION CORRELATION COEFFICIENT, ρ

Example

We will use the data and calculations of the previous section, and set out the seven-step method:

1. H_0: $\rho = 0$. This implies no correlation between the variables in the population.
2. H_1: $\rho > 0$. This implies that there is a positive correlation in the population, i.e. that increasing height is associated with increasing weight.
3. 5% significance level.
4. The calculated test statistic is:

$$Calc\ t = r\sqrt{\frac{n-2}{1-r^2}}$$

(Notice that this formula contains both r and n.)

$$= 0.874\sqrt{\frac{6-2}{1-0.874^2}} \quad \text{for the height/weight data,}$$
$$= 3.61.$$

5. *Tab t* = 2.132 from Table D.5, for $\alpha = 0.05$ (one-sided H_1), $\nu = (n-2) = 6 - 2 = 4$ (for this formula and these data, respectively). (The way to remember that the number of degrees of freedom is $(n-2)$ is that this term occurs in the formula for *Calc t*.)
6. Since *Calc t* > *Tab t*, reject H_0 (refer to section 10.10 if necessary).
7. There is significant positive correlation between height and weight. *Assumption*: Height and weight are separately normally distributed.

You should read the next section before trying Worksheet 13.

13.4 THE INTERPRETATION OF SIGNIFICANT AND NON-SIGNIFICANT CORRELATION COEFFICIENTS

The following points should be considered whenever we try to interpret correlation coefficients.

First, a significant value of r (i.e. when the null hypothesis H_0: $\rho = 0$ is rejected) does not necessarily imply cause and effect. For the height/weight data it clearly makes little sense to talk about 'height causing weight' or vice versa, but it might be reasonable to suggest that both variables are caused by (meaning 'depend on') a number of variables such as sex, heredity, diet, exercise and so on.

For the kinds of example quoted regularly by the media, we must be equally vigilant. Claims such as 'eating animal fats causes heart disease', 'wearing a certain brand of perfume causes a person to be more sexually attractive', 'reducing inflation causes a reduction in unemployment' may or may not be true. They are virtually impossible to substantiate without controlling or allowing for many other factors which may influence the chances of getting heart disease, the level of sexual attraction and the level of unemployment, respectively. Such careful research is difficult, expensive and time-consuming, even in cases where the other factors may be controlled or allowed for. Where they may not be, it is misleading to draw confident conclusions.

Second, the correlation coefficient measures the *linear* association between the variables. So a scatter diagram may indicate non-linear correlation but the Pearson correlation coefficient may have a value close to zero. For example a random sample of ten runners taking part in a local 'fun-run' of 10 miles may give rise to a scatter diagram such as Fig. 13.3, if the time to complete the course is plotted against the age of the runner. A clear curvilinear relationship exists but the value of r would be close to zero.

Third, a few outlying points may have a disproportionate effect on the value of r, as in Fig. 13.4. In Fig. 13.4(a) the inclusion of the **outlier** would give a smaller value of r than if it were excluded from the calculation. In Fig. 13.4(b) the inclusion of the outlier would give a larger value of r.

In fact, in both cases the assumption that both variables are normally distributed looks suspect. In Fig. 13.4(a) the outlier has a value of y which is far away from the other values, and in Fig. 13.4(b) both the x and the y values of the outlier are extreme. However, we must not discard outliers simply because they do not fit into the pattern of the other points. We are justified in suspecting that some mistake may have

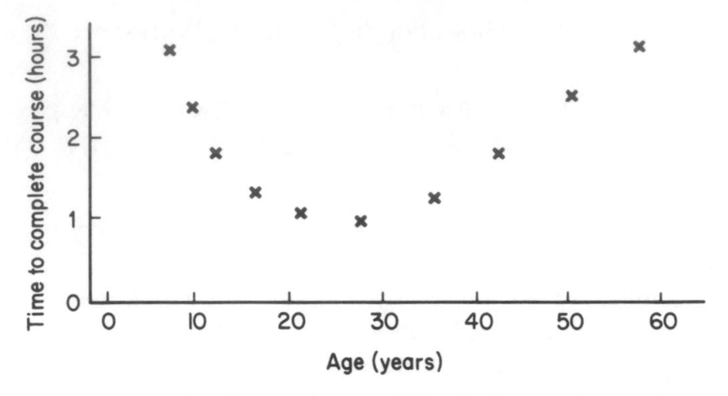

Fig. 13.3 Scatter diagram for time to complete course and age of runner for a sample of ten.

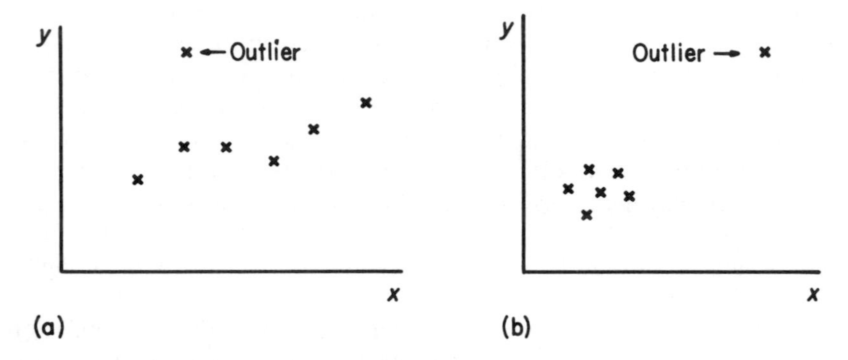

Fig. 13.4 Two scatter diagrams, each with an 'outlier'.

been made in calculating the x and y values, or in plotting the point, or in some other way.

Fourth, the value of r may be restricted, and be non-significant in a hypothesis test, because the ranges of the x and y variables are restricted. For example, the value of r between students' A-level count and their subsequent performance in a particular college of higher education may be restricted by the fact that, for that college, only students with A-level counts above a certain minimum are admitted while students with A-level counts which are very high may choose to go to other colleges (higher up in the 'pecking order'). The value of r for students actually admitted to a particular college may be lower than it would be if entry were unrestricted (Fig. 13.5).

Fifth and finally, nonsense correlations may result if two variables have increased or decreased in step over a period of time, but common sense

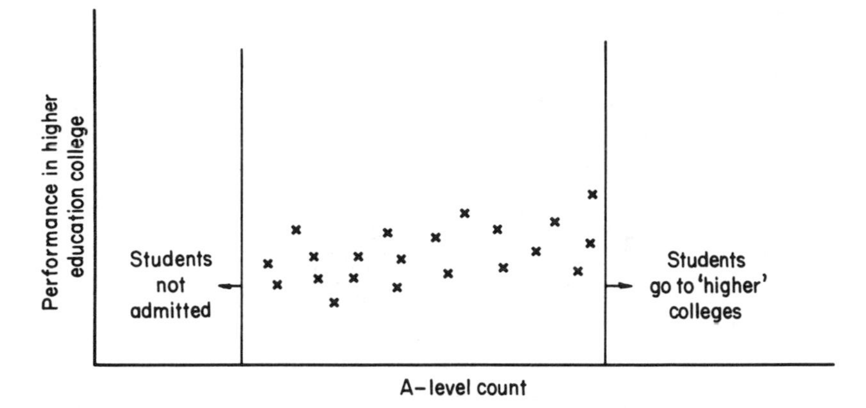

Fig. 13.5 Scatter diagram for A-level count and performance in higher education college.

indicates that the variables are clearly unconnected. There are many examples of this type of correlation: the number of violent crimes committed and doctors' salaries have increased over the last 10 years, and the correlation coefficient (for the 10 'individual' years) may be significant. Clearly it would be nonsense to conclude that giving more money to doctors increases the incidence of violent crime. Another nice example is the observation made in a Swedish town that in years when relatively more storks built their nests on house chimneys, relatively more babies were born in the town, and vice versa. A scatter diagram would have shown a possibly significant correlation, hence the idea that storks bring babies was born.

It may occur to you that, with all the above reservations, there is little to be gained by calculating the value of a correlation coefficient and testing it for significance. The interpretation we can place on a significant value of r is that 'such a value is unlikely to have arisen by chance if there is no correlation in the population, so it is reasonable to conclude that there is some correlation in the population'. To extend this conclusion to one of cause and effect, for example, requires much more information about other possible causal variables and consideration of the points made above in this section.

13.5 SPEARMAN'S RANK CORRELATION COEFFICIENT

If the two quantitative variables of interest are not normally distributed, Spearman's rank correlation coefficient may be calculated by ranking the sample data, separately for each variable, and using the formula:

$$r_s = 1 - \frac{6\Sigma d^2}{n^3 - n}$$

where r_s is the symbol for the sample value of Spearman's coefficient of rank correlation, and Σd^2 means the sum of the squares of the differences in the ranks of the n individuals. A non-parametric hypothesis test may then be carried out.

The formula above applies only when there are no 'tied ranks'. A 'tie' occurs when two (or more) sample values of either variable are equal, and so are given the same rank. The calculation of Spearman's r_s in the case of tied ranks is discussed in section 13.7.

Example (with no tied ranks)

For comparison purposes, the same data will be used as for the Pearson's r example of section 13.2. The data are repeated in Table 13.2, which also demonstrates the calculation of Σd^2. We calculate r_s as follows:

$$r_s = 1 - \frac{6 \times 2}{6^3 - 6}$$
$$= 1 - 0.057$$
$$= 0.943.$$

What does a sample value of 0.943 for r_s tell us? The possible range of values for r_s is -1 to $+1$ (the same as the possible range for Pearson's r). If $r_s = +1$, there is perfect agreement in the rankings. If $r_s = -1$, there is perfect disagreement (the highest rank for one variable corresponding to the lowest rank for the other variable, and so on). If $r_s = 0$, a particular rank for one variable may correspond with any rank for the other variable. So a value of $r_s = 0.943$ is high positive correlation (as we found for the same data when we calculated Pearson's r). Once again, though,

Table 13.2 Heights (cm) and weights (kg) ranked for a sample of six students

Height	*Weight*	*Rank of height*	*Rank of weight*	d^2
170	57	1	1	0
175	64	2	2	0
176	70	3	3	0
178	76	4	5	1
183	71	5	4	1
185	82	6	6	0
				$\Sigma d^2 = 2$

this value should not be judged in isolation, since we must also take into account the number of 'individuals' n, which we do by carrying out a formal hypothesis test, which we now describe.

13.6 HYPOTHESIS TEST FOR SPEARMAN'S RANK CORRELATION COEFFICIENT

Example

Using the sample data from the example in the previous section:

1. H_0: The ranks of height and weight are uncorrelated.
2. H_1: High ranks of height correspond to high ranks of weight (one-sided alternative).
3. 5% significance level.
4. *Calc* $r_s = 0.943$, from previous section.
5. *Tab* $r_s = 0.829$, from Table D.10 of Appendix D for $n = 6$, one-sided H_1, and 5% significance level.
6. Since *Calc* $r_s >$ *Tab* r_s, we reject H_0.
7. There is significant positive correlation between the ranks of height and weight (5% level).
   Assumption: We must be able to rank each variable.

The extensive notes in section 13.4 on the interpretation of correlation coefficients apply equally to both the Pearson and the Spearman coefficients.

13.7 SPEARMAN'S RANK CORRELATION COEFFICIENT IN THE CASE OF TIES

In section 13.5 it was stated that the formula for r_s did not apply in the case of tied ranks. In this situation, we can either introduce a modified and more complicated formula for r_s or we can use the following ingenious method. In the case of tied ranks, Pearson's r is calculated using the ranks rather than the original observed values of the variables. It can be shown that the resulting value is the correct value of Spearman's r_s, and this can then be tested for significance as in section 13.6.

Example

A random sample of ten students were asked to rate two courses they had all taken on a ten-point scale. A rating of 1 means 'absolutely dreadful', while a rating of 10 means 'absolutely wonderful'. The data are given in the first two columns of Table 13.3. Here we are not interested in whether one course has a higher mean rating than the other (in which case a Wilcoxon signed rank test would be appropriate – although unnecessary here since the sample mean ratings are equal), but we are interested in whether there is a significant correlation between the ratings. In other words, do students who rate one course highly, relative to the ratings of other students, also tend to rate the other course highly, relative to the ratings of other students, and vice versa? The scatter diagram of Fig. 13.6 indicates that the correlation may be positive but small, and we will hopefully confirm this subjective judgement when we calculate the sample value of the Spearman rank correlation coefficient.

For the ranks in Table 13.3, where x represents the ranks for the

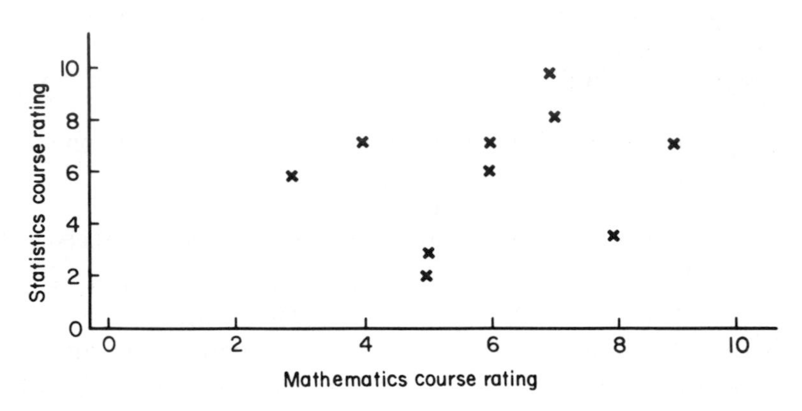

Fig. 13.6 Scatter diagram for the ratings of ten students taking courses in statistics and mathematics.

Table 13.3 The ranks of the statistics and mathematics course ratings of ten students

Statistics rating	*Mathematics rating*	*Ranks of statistics rating* (x)	*Ranks of mathematics rating* (y)
7	6	7	5.5
6	6	4.5	5.5
3	5	2	3.5
8	7	9	7.5
2	5	1	3.5
6	3	4.5	1
7	9	7	10
7	4	7	2
10	7	10	7.5
4	8	3	9

statistics course and y the ranks for mathematics, we have the following summary values:

$\Sigma x = 55$, $\Sigma x^2 = 382.5$, $\Sigma y = 55$, $\Sigma y^2 = 383.5$, $\Sigma xy = 331.75$, $n = 10$,

Hence Pearson's r is given by

$$r = \frac{331.75 - \frac{55 \times 55}{10}}{\sqrt{\left(382.5 - \frac{55^2}{10}\right)\left(383.5 - \frac{55^2}{10}\right)}}$$

$$= \frac{29.25}{\sqrt{80 \times 81}}$$

$$= 0.3634$$

Since *Tab* $r_s = 0.648$, for $n = 10$ and two-sided H_1, H_0 is not rejected and we conclude that the two sets of ratings are not significantly correlated (5% level).

13.8 CORRELATION COEFFICIENTS USING MINITAB

Minitab can give a scatter diagram using the command PLOT, and Pearson's r using the command CORRELATION. However, the test for the significance of r is not available, nor is Spearman's r_s unless the data are first ranked and the method of the previous section is used. The first correlation command in Table 13.4 gives Pearson's r, the second gives Spearman's r_s,

Table 13.4 Mintab commands for scatter diagram and correlation coefficients for heights and weights of six adults

```
MTB> SET C1
DATA> 170  175  176  178  183  185
DATA> END
MTB> SET C2
DATA> 57  64  70  76  71  82
DATA> END
MTB> NAME  C1  'HEIGHT'
MTB> NAME  C2  'WEIGHT'
PLOT> 'HEIGHT' 'WEIGHT'
MTB> CORRELATION  'HEIGHT'  'WEIGHT'
MTB> RANK  C1  C3
MTB> RANK  C2  C4
MTB> CORRELATION C3 C4
```

13.9 SUMMARY

As in Chapter 12 inferences from bivariate sample data are discussed, but in this chapter the case in which the variables are quantitative (rather than categorical) is covered.

The scatter diagram is a useful and important summary of this type of data. A measure of the degree of association between the variables is provided by a correlation coefficient. If both variables are normally distributed, Pearson's r is the appropriate coefficient. In other cases we may use Spearman's rank correlation coefficient, assuming the data are capable of being ranked.

Hypothesis tests may be used to test the significance of both coefficients, Pearson's test being more powerful if both variables are 'normal'.

There are several important points to bear in mind when we try to interpret correlation coefficients.

WORKSHEET 13: CORRELATION OF QUANTITATIVE VARIABLES

Fill in the gaps in Questions 1–6.

1. If two quantitative variables are measured for a number of individuals the data may be plotted in a _____ _____.
2. A _____ _____ is a measure of the degree of association between two quantitative variables.
3. If it is reasonable to assume that each variable is normally distributed and we wish to obtain a measure of the degree of linear association between them, the appropriate _____ _____ to calculate is named _____'s and has the symbol _____. For the population the symbol is _____.
4. In calculating _____ we must decide which of our variables is the x variable and which is the y variable. However, the choice is _____.
5. The value of r (or r_s) must lie somewhere in the range _____ to _____. If the points on the scatter diagram indicate that as one variable increases the other variable tends to decrease the value of r will be _____. If the points show no tendency either to increase together or to decrease together the value of r will be close to _____.
6. In order to decide whether there is a significant correlation between the two variables we carry out a hypothesis test for the population parameter _____ if the variables can be assumed to be _____ _____. If we cannot make this assumption the null hypothesis is that the rankings of the two variables are _____.

7. The percentage increase in unemployment and the percentage increase in manufacturing output were recorded for a random sample of ten industrialized countries over a period of a year. The data are recorded below. Draw a scatter diagram. Is there a significant correlation? What further conclusions can be drawn, if any?

Percentage increase in unemployment	*Percentage increase in manufacturing output*
10	−5
5	−10
20	−12
15	−8
12	−4
2	−5
−5	−2
14	−15
1	6
−4	5

8. A company owns eight large luxury hotels, one in each of eight geographical areas. Each area has a different commercial television channel. To estimate the effect of television advertising the company carried out a month's trial in which the number of times a commercial, advertising the local luxury hotel, was shown was varied from one area to another. The percentage increase in the receipts of each hotel over the three months following the month's trial was also calculated:

Area	1	2	3	4	5	6	7	8
Number of times the commercial shown	0	0	0	10	20	30	40	50
Percentage increase in receipts	−2	5	10	5	7	14	13	11

What conclusions can be drawn?

9. In a mountainous region a drainage system consists of a number of basins with rivers flowing through them. For a random sample of seven basins, the area of each basin and the total length of the rivers flowing through each basin are as follows:

Basin number	*Area (sq.km)*	*River length (km)*
1	7	10
2	8	8
3	9	14
4	16	20
5	12	11
6	14	16
7	20	10

Are larger areas associated with longer river lengths?

10. From the data in the table that follows, showing the percentage of the population of a country using filtered water and the death rate due to typhoid for various years, calculate the correlation coefficient and test its significance at the 5% level. What conclusions would you draw about the cause of the reduction in the typhoid death rate from 1900 to 1912?

Year	*Percentage using filtered water*	*Typhoid death rate per 100 000 living*
1900	9	36
1902	12	37
1904	16	35
1906	21	32
1908	23	27
1910	35	22
1912	45	14

11. A random sample of 20 families had the following annual income and annual savings in thousands of pounds:

Income	*Savings*
5.1	0.2
20.3	0.5
25.2	0.3
15.0	5.7
10.3	0.7
15.6	1.3
16.0	0.4
7.3	4.2
8.6	2.0

Income	*Savings*
12.3	0.6
14.0	0.3
8.9	0.1
12.4	0.0
14.0	0.7
16.0	0.2
14.0	0.3
15.3	1.0
12.4	0.6
10.3	0.5
11.3	0.7

Is there a significant positive correlation between income and savings?

12. For the data in Appendix A, which are the only two continuous variables? Which correlation coefficient should be calculated to measure the degree of association between them? Obtain this coefficient for the first 10 students only, and test its significance at the 5% level.

14

Regression analysis

14.1 INTRODUCTION

When two quantitative variables are measured for a number of individuals we may be not so much interested in a measure of association between the variables (provided by a correlation coefficient) as in predicting the value of one variable from the value of the other variable. For example, if trainee salesmen take a test at the end of their training period, can the test score be used to predict the first-year sales, and how accurate is the prediction? One way to answer such a question is to collect both the test score and the first-year sales of a number of salesmen and from these sample data develop an equation relating these two variables. This equation is an example of a **regression equation**, the simplest type of which is a **simple linear regression equation** which can be represented by a straight line on the scatter diagram for the two variables.

A couple of observations are worth making here. First, the word 'regression' is one for which the original meaning is no longer useful. In the nineteenth century, Galton collected the heights of fathers and their sons and put forward the idea that, since very tall fathers tended to have slightly shorter sons, and very short fathers tended to have slightly taller sons, there would be what Galton called a 'regression to the mean'. Second, the word 'simple' here implies that we are using one variable (rather than many as in 'multiple' regression) to predict another variable. Since this is the only case we will consider, the word 'simple' will not be used again to describe regression equations.

Before we consider how to obtain a linear regression equation, we should draw a scatter diagram first to decide whether the relationship between the variables appears to be reasonably linear (this was not the case in Fig. 13.3, for example).

For the example above of the salesmen, the linear regression equation will be of the form

$$(\text{first-year sales}) = a + b \times (\text{test score})$$

where a and b are values we can calculate from the sample data.

In general, if we call the variable we wish to predict the y variable and the variable we wish to do the predicting the x variable, the linear regression equation for 'y on x' is:

$$y = a + bx$$

The symbol b represents the **slope** (or **gradient**) of the line, and is also sometimes called the **regression coefficient**, and the symbol a represents the **intercept** (i.e. the value of y where the line crosses the y axis).

14.2 DETERMINING THE REGRESSION EQUATION FROM SAMPLE DATA

Example

Suppose that for a random sample of eight salesmen their first-year sales and test scores as trainees are as shown in Table 14.1. Notice that we have labelled sales as the y variable since we wish to predict it; and test score as the x variable, since it will be used to make a prediction.

The scatter diagram of the eight points (Fig. 14.1) shows that there appears to be a reasonable linear relationship between the variables. We will now calculate a and b (in reverse order) using the formulae:

$$b = \frac{\Sigma xy - \dfrac{\Sigma x \Sigma y}{n}}{\Sigma x^2 - \dfrac{(\Sigma x)^2}{n}} \quad \text{and} \quad a = \bar{y} - b\bar{x},$$

where $\bar{x}$ and $\bar{y}$ are the sample means of x and y, so

$$\bar{x} = \frac{\Sigma x}{n}, \quad \bar{y} = \frac{\Sigma y}{n}$$

Table 14.1 First-year sales (£ thousands) and test scores of eight salesmen

First-year sales y	*Test score* x
105	45
120	75
160	85
155	65
70	50
150	70
185	80
130	55

and n is the number of 'individuals'. For the data in Table 14.1,

$$\Sigma x = 525 \qquad \Sigma x^2 = 35\,925$$
$$\Sigma y = 1075 \qquad \Sigma y^2 = 153\,575$$
$$\Sigma xy = 73\,350 \qquad n = 8$$

(refer to similar calculations in section 13.2 if necessary). Using these values, we obtain

$$b = \frac{73\,350 - \dfrac{525 \times 1075}{8}}{35\,925 - \dfrac{525^2}{8}} = \frac{2803}{1472} = 1.904$$

$$a = \frac{1075}{8} - 1.904 \times \frac{525}{8} = 134.4 - 125.0 = 9.4.$$

The regression equation of y (first-year score) on x (test score) is:

$$y = 9.4 + 1.904x, \quad \text{or}$$
$$\text{(first-year sales)} = 9.4 + 1.904 \times \text{(test score)}$$

14.3 PLOTTING THE REGRESSION LINE ON THE SCATTER DIAGRAM

We can now plot the regression line for the example in the previous section to represent the regression equation $y = 9.4 + 1.904x$. Since two points determine a straight line, we can substitute any two values of x into the equation, calculate the corresponding values of y, and thus we obtain the co-ordinates of two points on the regression line. It is a good idea to use the minimum and maximum sample data observations of x. (For the reason, see note (b) below on extrapolation.)

The minimum observed value of x in Table 14.1 is 45. When $x = 45$, the 'predicted' value of y is $9.4 + 1.904 \times 45 = 95.1$. The maximum value of x is 85. When $x = 85$, the 'predicted' value of y is $9.4 + 1.904 \times 85 = 171.2$. We join the points (45, 95.1) and (85, 171.2) by a straight line on the scatter diagram, as shown in Fig. 14.1.

Notes

(a) The regression line is often referred to as the line of 'best fit'. In what sense is it best? The answer is that it is the line for which the sum of squares of the distances from the points of the scatter diagram to the line in the y direction is minimized. There are good theoretical reasons for using this criterion. The formulae for a and b which we used in section 14.2 were derived by calculus using this criterion, but the derivation of these formulae is beyond the scope of this book.

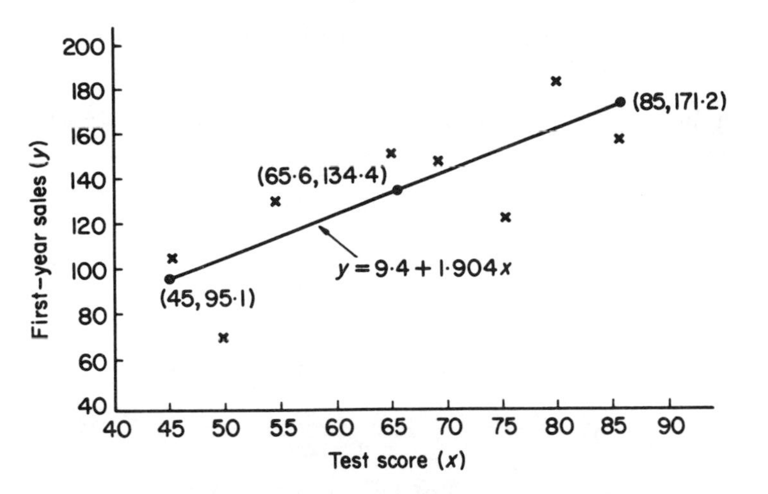

Fig. 14.1 Scatter diagram for first-year sales and test score.

(b) The regression line and the regression equation apply only within the range of the observed x values in the data. The regression line should not be extrapolated (extended) below the minimum value of x or above the maximum value of x. Nor should we use the regression equation for values of x outside the range of the sample data.

(c) As a check on the position of the line on the scatter diagram, it can be shown that the regression line should pass through the point $(\bar{x}, \bar{y})$. For the example this point (shown in Fig. 14.1) is (65.6, 134.4).

14.4 PREDICTING VALUES OF y

For a particular value of x (within the range of the sample data) the corresponding point on the regression line gives the predicted value of y. This may also be obtained, and with more accuracy, by substituting the particular value of x into the regression equation.

Example

Predict first-year sales for a test score of 60.

For $x = 60$, 'predicted' $y = 9.4 + 1.904 \times 60 = 123.6$. What does this mean? It is an estimate of the *mean* first-year sales for salesmen with test scores of 60.

Using the ideas of Chapter 9 we may also calculate a confidence interval for the mean sales we would predict for test scores of 60 (and any other test scores within the range of the sample data) to give a measure of the precision of the estimate of first-year sales for test scores of 60. These calculations are shown in the next section.

14.5 CONFIDENCE INTERVALS FOR PREDICTED VALUES OF y

The formula for a 95% confidence interval for the predicted value of y at some value $x = x_0$ is:

$$(a + bx_0) \pm ts_r\sqrt{\frac{1}{n} + \frac{(x_0 - \bar{x})^2}{\Sigma x^2 - \frac{(\Sigma x)^2}{n}}}$$

where

$$s_r^2 = \frac{\left[\Sigma y^2 - \frac{(\Sigma y)^2}{n}\right] - b^2\left[\Sigma x^2 - \frac{(\Sigma x)^2}{n}\right]}{n - 2}$$

and t is obtained from Table D.5 for $\alpha = 0.025$, $v = n - 2$. In using this confidence interval formula, the following assumptions are made:

(a) The data points are distributed approximately normally about the regression line in the y direction.
(b) The distribution is the same for all values of x.

The distribution referred to is shown graphically in Fig. 14.2. The assumption of normality is less important the larger the number of individuals, n. The assumption of a constant normal distribution requires that the variability about the line in the y direction is the same all along the line, and does not, for example, tend to increase (or decrease) significantly as x increases.

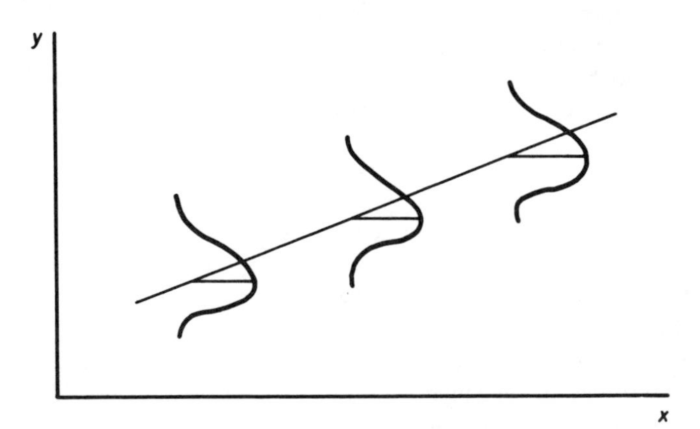

Fig. 14.2 Assumptions required in using the formula for confidence intervals for predicted y.

Example

For the example of the salesmen, we can use the confidence interval formula for values of x between 45 and 85.

$$s_r^2 = \frac{\left[153\,375 - \dfrac{1075^2}{8}\right] - 1.904^2\left[35\,925 - \dfrac{525^2}{8}\right]}{8 - 2}$$

$$= \frac{9122 - 1.904^2 \times 1472}{6}$$

$$= \frac{3786}{6}$$

$$= 631,$$

$$s_r = 25.1.$$

Notes

(a) The numerator of the formula for s_r^2 represents the sum of squares of the distances from the points to the line in the y direction. This sum of squares is 3786 for the data in the example. For any other line drawn on the scatter diagram the sum of squares will exceed 3786.

(b) s_r is called the *residual standard deviation*, and the distances from the points to the line are called *residuals* (Fig. 14.3).

At $x_0 = 60$, the 95% confidence interval for predicted y is

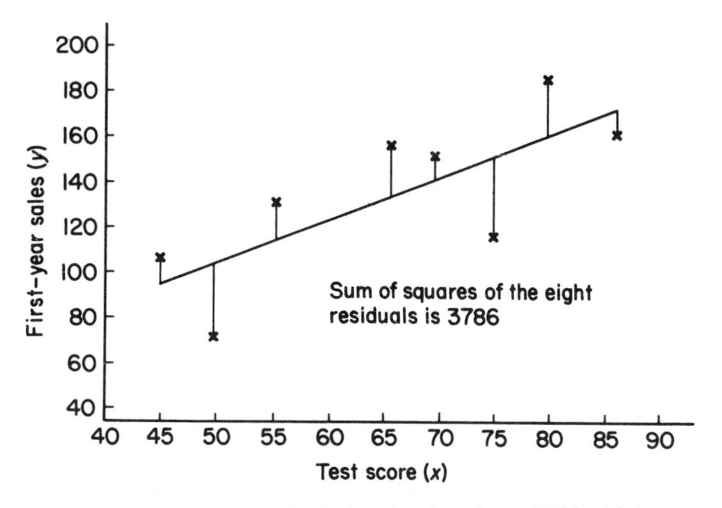

Fig. 14.3 The residuals for the data from Table 14.1.

$$(9.4 + 1.904 \times 60) \pm 2.447 \times 25.1\sqrt{\frac{1}{8} + \frac{(60 - 65.6)^2}{35\,925 - \dfrac{525^2}{8}}}$$

$$123.6 \pm 23.5$$

$$100.1 \text{ to } 147.1.$$

For a test score of 60, then, the mean first-year sales will be between 100.1 and 147.1 with 95% confidence. For other test scores, similar calculations can be made; in Fig. 14.4 the two curves above and below the regression line represent the 'locus' of all the 95% confidence intervals between test scores of 45 and 85. Notice how the 95% confidence interval is narrowest at $x = 65.6$, the mean value of x, and widens as we move away from the mean in either direction (increasing or decreasing x). This result agrees with the intuitive idea that we expect more precision in our prediction the nearer we are to the middle of the data, and conversely we expect less precision when we move towards the extremes of the data.

14.6 HYPOTHESIS TEST FOR THE SLOPE OF THE REGRESSION LINE

For the example we calculated $b = 1.904$ (in section 14.2). This is the slope of the regression line for the sample data. It is our estimate of the increase in y (first-year sales) for unit increase in x (test score).

We can also conceive of a regression line for a population of salesmen with an equation:

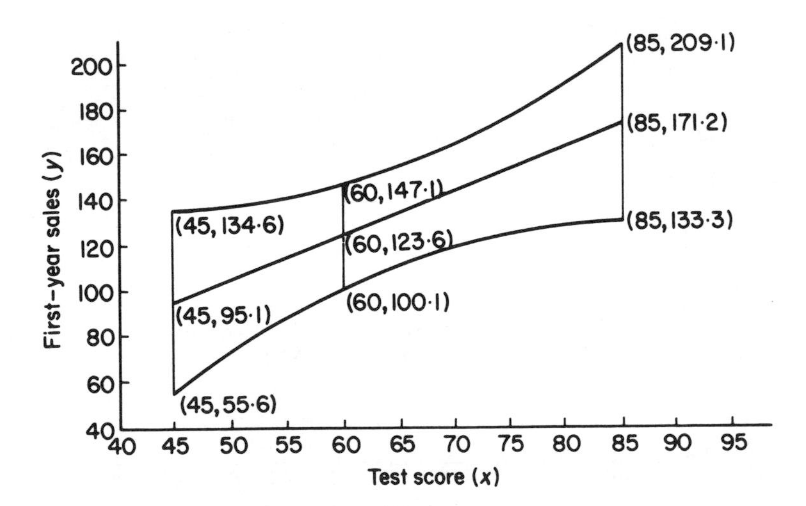

Fig. 14.4 The locus of 95% confidence intervals for predicted y, using data from Table 14.1.

$$y = \alpha + \beta x$$

where β is the slope, and α is the intercept. So our estimate of β is the sample estimate provided by b, namely 1.904. Could such a sample value have arisen if the population value was really zero (implying a horizontal population regression line)? To answer this question we carry out a hypothesis test as follows:

1. H_0: $\beta = 0$.
2. H_1: $\beta > 0$, one-sided, implying that higher test scores result in higher first-year sales.
3. 5% significance level.
4. The calculated test statistic for this test is:

$$Calc\ t = \frac{b}{s_r \Big/ \sqrt{\Sigma x^2 - \dfrac{(\Sigma x)^2}{n}}} = \frac{1.904}{25.1 \Big/ \sqrt{35\,925 - \dfrac{525^2}{8}}}$$

$$= 2.91$$

5. $Tab\ t = 1.943$, for $\alpha = 0.05/1 = 0.05$, $\nu = (n - 2) = 6$.
6. Since $Calc\ t > Tab\ t$, reject H_0.
7. The slope of the regression line is significantly greater than zero (5% level).

   Assumptions: The same assumptions apply as in section 14.5.

14.7 THE CONNECTION BETWEEN REGRESSION AND CORRELATION

In the previous section, if we had not rejected the null hypothesis H_0: $\beta = 0$, then it would have been reasonable to regard the population regression line as being horizontal, in other words parallel to the x axis, so that, whatever the value of x, the predicted value of y would have been the same. Another way of expressing the same idea is to describe the variables x and y as uncorrelated in this case.

On the other hand, if H_0: $\beta = 0$ is rejected in favour of H_1: $\beta > 0$ then higher values of x would give rise to higher predicted values of y, and we would say that the variables were positively correlated.

This intuitive reasoning is supported by the fact that a hypothesis test for Pearson's r (section 13.3) would result in values of *Calc t* and *Tab t* identical to those found in section 14.6. This statement is left for you to confirm by calculation.

14.8 REGRESSION ANALYSIS USING MINITAB

Try the Minitab commands shown in Table 14.2 which use the data in Table 14.1.

Table 14.2 Minitab commands for a regression analysis of the data in Table 14.1

```
MTB> SET  C1
DATA> 105  120  160  155  70  150  185  130
DATA> END
MTB> SET  C2
DATA> 45  75  85  65  50  70  80  55
DATA> END
MTB> NAME  C1  'SALES'
MTB> NAME  C2  'SCORE'
MTB> REGRESSION  'SALES'  1  'SCORE';
SUBC> PREDICT  45;
SUBC> PREDICT  60;
SUBC> PREDICT  85;
SUBC> RESIDUALS  C3.
MTB> PRINT  C1  C2  C3
```

Part of the output will be:

```
THE REGRESSION EQUATION IS
SALES = 9.4 + 1.904 SCORE
FIT        95% C.I.
95.1       (55.6, 134.6)
123.6      (100.1, 147.1)
171.2      (133.3, 209.1)
```

Notes

(a) The regression equation agrees with that given in section 14.2 for the same data.

(b) The predicted values of sales for test scores of 45, 60 and 85 are given in the column headed FIT along with the corresponding confidence intervals. The numbers shown agree with those given in sections 14.4, 14.5 and Fig. 14.4.

(c) The residuals are displayed in C3 (see also section 14.5).

(d) Minitab will also produce other output which is beyond the scope of this book, such as an analysis of variance table. It is worth pointing out, however, that if the '*P* value' is less than 0.05, then we can reject the null hypothesis that $\beta = 0$, in favour of a two-sided alternative in a test similar to that carried out in section 14.6.

14.9 TRANSFORMATIONS TO PRODUCE LINEAR RELATIONSHIPS

When the scatter diagram shows a non-linear pattern it may be possible to produce a linear pattern by *transforming* one or both of the variables.

Example

The annual income (in thousands of pounds) and the annual expenditure on meat (in hundreds of pounds) by a random sample of ten families of the same size is shown in Table 14.3.

The x and y variables are as in the table, and a new variable z has been calculated by taking logarithms to base 10 of the y values. The reason for introducing the new variable z is clear if we study the two scatter diagrams in Fig. 14.5. The plot of y against x has a definite tendency to be

Table 14.3 Income (£ thousands) and expenditure on meat (£ hundreds) for ten families

Annual income y	*Annual expenditure on meat* x	$z = \log_{10} y$
5	2	0.70
6	4	0.78
7	5	0.85
8	6	0.90
10	7	1.00
12	8	1.08
15	9	1.18
20	10	1.30
30	11	1.48
40	12	1.60

Fig. 14.5 Scatter diagrams for: (a) income versus expenditure on meat; (b) log income versus expenditure on meat.

non-linear, but the plot of z against x shows a stronger linear relationship between this pair of variables.

If we wish to predict income from expenditure on meat, this may be achieved by calculating the regression equation of 'z on x', and then transforming back using $z = \log_{10} y$.

Assuming a regression equation $z = a + bx$, and using the summations

$$\Sigma x = 74, \quad \Sigma x^2 = 640, \quad \Sigma z = 10.87, \quad \Sigma xz = 88.91, \quad n = 10,$$

we can calculate that $b = 0.0917$ and $a = 0.4085$. So $z = 0.4085 + 0.0917x$, which we transform back to read:

$$\log_{10} y = 0.4085 + 0.0917x.$$

This equation may be used to predict y from x. The equation may be represented as a straight line on the plot of z against x, but will be a curve on the plot of y against x.

The choice of what transformation to use in a particular case is beyond the scope of this book. Apart from the log transformation, other transformations (which may be appropriate depending on the type of data collected) are square roots (for Poisson data) or arcsines (for binomial data).

14.10 SUMMARY

(Simple) linear regression analysis is a method of deriving an equation relating two quantitative variables. The equation, which may be represented by a straight line on a scatter diagram, is:

$$y = a + bx,$$

where

$$b = \frac{\Sigma xy - \dfrac{\Sigma x \Sigma y}{n}}{\Sigma x^2 - \dfrac{(\Sigma x)^2}{n}} \quad \text{and} \quad a = \bar{y} - b\bar{x}$$

The equation may be used to predict values of y for given values of x within the range of the sample data.

Under certain assumptions, confidence intervals for predicted values of y may be calculated, the null hypothesis that the slope of the population regression line, β, is zero may be tested, and a confidence interval for β may be calculated (see Worksheet 14, Question 8).

The connection between regression and correlation was discussed, as was an example of transforming one of the variables to 'linearize' a non-linear pattern in the scatter diagram.

WORKSHEET 14: REGRESSION ANALYSIS

Fill in the gaps in Questions 1 and 2.

1. The purpose of (simple) regression analysis is to ____ values of one variable for particular values of another variable. We call the variable whose values we wish to predict the ____ variable, and the other we call the ____ variable.

2. Using sample values of the two variables the ____ diagram is drawn. If this appears to show a linear relationship between the variables we calculate a and b for the linear ____ equation. This equation may be represented by a ____ ____ on the scatter diagram.

3. The regression line is also called the line of 'best fit' because it minimizes the sum of squares of the distances from the points of the scatter diagram to the line in the y direction. For the example in section 14.2, this sum of squares is 3786. Draw any other line 'by eye' on the scatter diagram which you think may be a better fit and calculate the sum of squares for your line. You should not be able to beat 3786, rounding errors excepted.

4. The following table gives the number of bathers at an open-air swimming pool and the maximum recorded temperature (°C) on ten Saturdays during one summer:

Number of bathers	*Maximum temperature*
290	19
340	23
360	20
410	24
350	21
420	26
330	20
450	25
350	22
400	29

Draw a scatter diagram and calculate the slope and intercept of the regression line which could be used to predict the number of bathers from the maximum temperature. Plot the regression line on the scatter diagram, checking that it passes through the point $(\bar{x}, \bar{y})$.

How many bathers would you predict if the forecast for the maximum temperature on the next Saturday in the summer was:

(a) 20°C, (b) 25°C, (c) 30°C?

Which of these predictions will be the least reliable?

5. In order to estimate the depth of water (in metres) beneath the keel of a boat, a sonar measuring device was fitted. The device was tested by observing the sonar readings over a number of known depths, and the following data were collected:

Sonar reading	0.15	0.91	1.85	3.14	4.05	4.95
True depth of water	0.2	1	2	3	4	5

Draw a scatter diagram for these data and derive a linear regression equation which could be used to predict the true depth of water from the sonar reading. Predict the true depth from a sonar reading of zero and obtain a 95% confidence interval for your prediction. Interpret your result.

6. The percentage moisture content of a raw material and the percentage relative humidity of the atmosphere in the store where the material is kept were measured on seven randomly selected days. On each day one randomly selected sample of raw material was used.

Relative humidity	30	35	52	38	40	34	60
Moisture	7	10	14	9	11	6	16

Draw a scatter diagram and derive a linear regression equation which could be used to predict the moisture content of the raw material from the relative humidity. Use the equation to predict moisture content for a relative humidity of:

(a) 0%, (b) 50%, (c) 100%.

Also test the hypothesis that the slope of the population regression line is zero.

7. The data below give the weight (kg) and daily food consumption (in hundreds of calories) for 12 obese adolescent girls. Calculate the best-fit linear regression equation which would enable you to predict food consumption from weight, checking initially that the relationship appears to be linear.

Weight	85	95	80	60	95	85	90	80	85	70	65	75
Food consumption	32	33	33	24	39	32	34	28	33	27	26	29

What food consumption would you predict, with 95% confidence, for adolescent girls weighing:

(a) 65, (b) 80, (c) 95 kg?

8. To see if there is a relationship between the size of boulders (cm) in a stream and the distance (km) from the source of the stream, samples of boulders were measured at 1 kilometre intervals. The average sizes of boulders found at various distances were as follows:

Distance downstream	1	2	3	4	5	6	7	8	9	10
Average boulder size	105	85	80	85	75	70	75	60	50	55

Find the regression equation which could be used to predict average boulder size from distance downstream. Plot the regression line on the scatter diagram. Test the null hypothesis that $\beta = 0$ against the alternative that $\beta < 0$. Also obtain a 95% confidence interval for β using the formula:

$$b \pm \frac{ts_r}{\sqrt{\Sigma x^2 - \dfrac{(\Sigma x)^2}{n}}}$$

where t is from Table D.5 for $\alpha = 0.025$, and $v = n - 2$.

9. The number of grams of a given salt which will dissolve in 100 g of water at different temperatures (°C) is shown below:

Temperature	0	10	20	30	40	50	60	70
Weight	53.5	59.5	65.2	70.6	75.5	80.2	85.5	90.0

Find the regression equation which could be used to predict weight of salt from temperature. Plot the regression line on the scatter diagram. Predict the weight of salt which you estimate would dissolve at temperatures of:

(a) 25°C (b) 55°C (c) 85°C

and comment on your results.

10. A random sample of ten people who regularly attempted the daily crossword puzzle in a particular national newspaper were asked to time themselves on a puzzle which none of them had seen before. Their times (in minutes) to complete the puzzle and their scores in a standard IQ test were as follows:

IQ	120	100	130	110	100	140	130	110	150	90
Times	9	7	13	8	4	5	16	7	5	13

What conclusions can be drawn from these data?

11. The following data show the values of two variables x and y obtained in a laboratory experiment on seven rats:

x	0.4	0.5	0.7	0.9	1.3	2.0	2.5
y	2.7	4.4	5.4	6.9	8.1	8.4	8.6

It is thought that there is a linear relationship between either (a) y and x or (b) y and $1/x$. Plot y against x, and y against $1/x$, and decide from the two scatter diagrams which will result in a better linear relationship. Calculate the coefficients of the appropriate linear regression equation and plot the regression line on the scatter diagram. Obtain the predicted value of y for $x = 1$ and calculate a 95% confidence interval for your prediction.

Without performing any further calculations, decide which of the following will be greater:

(i) The Pearson correlation coefficient between y and x.

(ii) The Pearson correlation coefficient between y and $1/x$.

Goodness-of-fit tests

15.1 INTRODUCTION

We return, in this the final chapter, to a one-variable problem, namely the problem of deciding whether our sample data for one variable could have been selected from a particular type of probability distribution. Four types of distribution will be considered:

Type of distribution	*Type of variable*
'Simple proportion'	Categorical
Binomial	Discrete
Poisson	Discrete
Normal	Continuous

In the first three cases a χ^2 test will be used to see how closely the observed frequencies of the sample data agree with the frequencies we would expect under the null hypothesis that sample data actually do come from the type of distibution being considered (refer to Chapter 12 now if you are unfamiliar with the test). In addition, a small-sample test, namely the Shapiro–Wilk test, will be discussed for the case of the normal distribution.

15.2 GOODNESS-OF-FIT FOR A 'SIMPLE PROPORTION' DISTRIBUTION

We define a simple proportion distribution as one for which we expect the frequencies of the various categories, into which the 'values' of a categorical variable will fall, to be in certain numerical proportions or ratios.

Example

The ratio of numbered cards to picture cards in a pack of 52 is 36 to 16, which we could write as 36:16. If we selected cards randomly with replacement we would expect the proportions of numbered cards and picture cards to be 36/52 and 16/52, respectively.

Example

Suppose that there is a genetic theory that adults should have hair colours of black, brown, fair and red in the ratios 5:3:1:1. If this theory is correct we expect the frequencies of black, brown, fair and red hair to be in the proportions:

$$\frac{5}{5+3+1+1}, \quad \frac{3}{5+3+1+1}, \quad \frac{1}{5+3+1+1}, \quad \frac{1}{5+3+1+1}$$

or 5/10, 3/10, 1/10 and 1/10.

If we take a random sample of 50 people to test this theory, we would expect 25, 15, 5 and 5 to have black, brown, fair and red hair, respectively (we simply multiplied the expected proportions by the sample size). We can then compare these expected frequencies with the frequencies we actually observed in the sample and calculate a χ^2 statistic.

It is convenient to set out this calculation in the form of a table (Table 15.1). Notice that the method of calculating the E values ensures that the sum of the E values equals the sum of the O values.

We now set out the seven-step hypothesis test for this example:

1. H_0: Genetic theory is correct. Sample data do come from a 5:3:1:1 distribution.
2. H_1: Genetic theory is not correct (two-sided).
3. 5% significance level.
4. From Table 15.1 we have

$$Calc\ \chi^2 = \sum \frac{(O-E)^2}{E} = 1.36$$

Table 15.1 Calculation of χ^2 for a 5:3:1:1 distribution

Hair colour	*Expected proportions*	*Expected frequencies (E)*	*Observed frequencies (O)*	$\frac{(O-E)^2}{E}$
Black	5/10	25	28	0.36
Brown	3/10	15	12	0.60
Fair	1/10	5	6	0.20
Red	1/10	5	4	0.20
		50	50	*Calc* χ^2 = 1.36

5. *Tab* $\chi^2 = 7.82$ for α = sig. level = 0.05 (even though H_1 is two-sided), and ν = (number of categories − 1) = 4 − 1 = 3, from Table D.9.
6. Since *Calc* $\chi^2 <$ *Tab* χ^2, do not reject H_0.
7. It is reasonable to conclude that the genetic theory is correct (5% level).

Notes

(a) The formula for *Calc* χ^2 is only valid if all the E values are all at least 5. If any E value is less than 5, it may be sensible to combine adjacent categories so that all E values for the new categories are all at least 5.

(b) The formula

$$Calc\ \chi^2 = \sum \frac{(O - E)^2}{E}$$

is used if $\nu > 1$. If $\nu = 1$ use

$$Calc\ \chi^2 = \sum \frac{(|O - E| - 1/2)^2}{E}$$

(applying Yates's correction as in section 12.3).

(c) The formula for degrees of freedom, ν = (number of categories − 1), may be justified by reference to section 9.7, the one restriction being that the sum of the E values must equal the sum of the O values. Only three of the E values may be determined independently in the example, so there are three degrees of freedom.

(d) If some categories are combined (see (a) above), the number of categories to be used to calculate the degrees of freedom is the number after combinations have been made.

15.3 GOODNESS-OF-FIT FOR A BINOMIAL DISTRIBUTION

Suppose we carry out n trials where each trial can result in one of only two possible outcomes, which we call 'success' and 'failure'. Suppose we repeat this set of n trials several times and observe the frequencies for the various numbers of successes which occur. We may then carry out a χ^2 test to decide whether it is reasonable to assume that the number of successes in n trials has a binomial distribution with a value for p which we can either estimate from the observed frequencies or sometimes specify without reference to the observed frequencies. (It will be assumed that you are familiar with the binomial distribution as described in Chapter 6.)

Example

In an experiment in extra-sensory perception (ESP) four cards marked A, B, C and D are used. The experimenter, unseen by the subject, shuffles the cards and selects one. The subject tries to decide which card has been selected. This procedure is repeated five times for each of a random sample of 50 subjects. The number of times, out of a maximum of five, that each subject correctly identifies a selected card is counted.

Suppose that the data for all 50 subjects are recorded in a table such as Table 15.2. Is there evidence that subjects are simply guessing? We may regard the testing of each subject as a set of five trials, each trial having one of two possible outcomes, 'correct decision' or 'incorrect decision'. This set of five trials is repeated (on different subjects) a total of 50 times. The second row in Table 15.2 gives the observed frequencies (O) for the various possible numbers of correct decisions.

If subjects are guessing then the probability of a correct decision is 1/4 or 0.25 for each selection, since the four cards are equally likely to be selected. So the question above, 'Is there evidence that subjects are simply guessing?', is equivalent to the question: 'Is it reasonable to suppose that the data in Table 15.2 come from a binomial distribution with $n = 5$, $p = 0.25$?'

The expected frequencies (E) for the various numbers of correct decisions are obtained by assuming, for the purposes of the calculation, that we are dealing with a binomial distribution with $n = 5$, $p = 0.25$. First we calculate the probabilities of 0, 1, 2, 3, 4 and 5 correct decisions (using the methods of Chapter 6). These probabilities are multiplied by the total of the observed frequencies (50 in the example) to give the expected frequencies. These calculations and the calculation of χ^2 are set out in Table 15.3.

We now set out the seven-step hypothesis test for this example:

1. H_0: Sample data do come from a $B(5, 0.25)$ distribution, implying that the subjects are guessing.
2. H_1: Sample data do not come from a $B(5, 0.25)$ distribution, which might imply that some subjects have powers of ESP.
3. 5% significance level.
4. *Calc* $\chi^2 = 5.80$, from Table 15.3.
5. *Tab* $\chi^2 = 7.82$, for $\alpha = 0.05$, $\nu =$ number of categories $- 1 = 3$. (Refer to notes (a) and (b) below and Table D.9.)

Table 15.2 Results of an ESP experiment, 50 subjects, 5 trials per subject

Number of correct decisions	0	1	2	3	4	5
Number of subjects	15	18	8	5	3	1

Table 15.3 Calculation of χ^2 for a binomial distribution

Number of correct decisions	P(*x*)	E = P(*x*) × *50*		*O*		$\frac{(O-E)^2}{E}$
0	0.2373	11.9		15		0.81
1	0.3955	19.8		18		0.16
2	0.2637	13.2		8		2.05
3	0.0879	4.4	}	5	}	
4	0.0146	0.7	5.2	3	9	2.78
5	0.0010	0.1	}	1	}	
Total	1.0000	50.1		50		*Calc* χ^2 = 5.80

Notes
(a) The probabilities P(x) were obtained from Table D.1 for $n = 5$, $p = 0.25$.
(b) Because three E values are less than 5, the bottom three categories have been combined.
(c) The totals of the E and the O columns are equal (apart from rounding errors).

6. Since *Calc* χ^2 < *Tab* χ^2, do not reject H_0.
7. It is reasonable to suppose that subjects are guessing (5% level).

Notes

(a) Although there were six categories initially, there were only four after combinations.
(b) In cases where the value of p is not specified by the experimental set-up, unlike the example above, we have to estimate it from the data of the observed frequencies (see Worksheet 15, Question 5), and we lose a further degree of freedom.

15.4 GOODNESS-OF-FIT FOR A POISSON DISTRIBUTION

Suppose that we observe the number of times a particular event occurs in each of a number of units of time (or space). Can we conclude that the number of occurrences of the event per unit time (or space) has a Possion distribution, implying randomly occuring events?

Example

Suppose that the number of major earthquakes occurring per month is collected for 100 months, as in Table 15.4.

Table 15.4 Number of earthquakes occurring in 100 months

Number of earthquakes per month	0	1	2	3	4
Number of months	57	31	8	3	1

The observed frequencies (O) for the various numbers of earthquakes per month are given in the second row of the table. The expected frequencies (E) for the various numbers of earthquakes are obtained by assuming, for the purposes of the calculation, that we are dealing with a Poisson distribution. The parameter m, the mean of the distribution, is estimated by the sample mean number of earthquakes per month:

$$m = \frac{\text{total number of earthquakes}}{\text{total number of months}}$$

$$= \frac{57 \times 0 + 31 \times 1 + 8 \times 2 + 3 \times 3 + 1 \times 4}{100}$$

$$= 0.6$$

For the Poisson distribution with a mean $m = 0.6$, we can obtain the probabilities of 0, 1, 2, 3 and 4 or more earthquakes (using the methods of Chapter 6). These probabilities are multiplied by the total of the observed frequencies (100 in the example) to give the expected frequencies.

These calculations and the calculation of the χ^2 test statistic are set out in Table 15.5. We now set out the seven-step hypothesis test for this example:

1. H_0: Sample data come from a Poisson distribution, implying earthquakes occur randomly in time.
2. H_1: Sample data do not come from a Poisson distribution.
3. 5% significance level.
4. *Calc* $\chi^2 = 0.19$, from Table 15.5.
5. *Tab* $\chi^2 = 3.84$ for $\alpha = 0.05$, $\nu =$ number of categories $- 1 - 1 = 1$ (see note below).
6. Since *Calc* $\chi^2 <$ *Tab* χ^2, do not reject H_0.
7. It is reasonable to assume a Poisson distribution, and that earthquakes occur randomly in time (5% level).

Notes

(a) There are only three categories after combinations. One degree of freedom is lost because of the restriction $\Sigma E = \Sigma O$, and another is lost because the parameter m is estimated from the sample data.

(b) Strictly speaking we should have used Yates's correction in calculating

Table 15.5 Calculation of χ^2 for a Poisson distribution

Number of earthquakes per month (x)	P(*x*)	*E* = P(*x*) × *100*		*O*		$\frac{(O-E)^2}{E}$
0	0.5488	54.9		57		0.08
1	0.3293	32.9		31		0.11
2	0.0988	9.9		8		
3	0.0197	2.0	12.2	3	12	0.00
4 or more	0.0034	0.3		1		
Total	1.0000	100.0		100		*Calc* χ^2 = 0.19

Notes

(a) The probabilities P(x) were obtained from Table D.2 for m = 0.6 (see section 6.12).
(b) The probability of '4 or more' (rather than '4') is calculated to ensure that the totals of the E and the O columns are equal, apart from rounding errors.
(c) The bottom three categories have been combined because two E values are less than 5.

χ^2. However, this would have the effect of reducing *Calc* χ^2, and so the null hypothesis would still not have been rejected.

15.5 THE SHAPIRO–WILK TEST FOR NORMALITY

Although it is possible to use a χ^2 test to test for normality if we have a sample of at least 50, the assumption of normality required to carry out most hypothesis tests (for example, the paired and unpaired t tests of Chapter 10):

(a) only requires the assumption of approximate normality,
(b) is less important if the sample size is large.

The problem of normality is really only important in most situations for sample sizes below n = 15, as we pointed out in section 9.4. There we used dotplots and judgement. A more objective method for small samples is provided by the Shapiro–Wilk test as follows.

Rank the n sample observations in increasing order, referring to them as $x_{(1)}, x_{(2)}, \ldots, x_{(n-1)}, x_{(n)}$, where $x_{(1)}$ means the smallest observed value, $x_{(2)}$ the next smallest and so on, so that $x_{(n)}$ is the largest. Then calculate

$$b = a_1(x_{(n)} - x_{(1)}) + a_2(x_{(n-1)} - x_{(2)}) + \cdots$$

where $a_1, a_2, \ldots$ are coefficients taken from Table D.11. Now calculate the test statistic *Calc W*, given by the formula

$$Calc\ W = \frac{b^2}{(n-1)s^2}$$

where s is the standard deviation of the n sample observations. Then look up *Tab W* in Table D.12. If *Calc W* > *Tab W*, we do not reject normality and hence we conclude that the data are consistent with the hypothesis of a normal distribution.

Example (n *even*)

Test the normality of the following sample of 10 sample observations (data from section 11.2):

150 500 240 120 130 300 140 160 110 200

In rank order these are:

110 120 130 140 150 160 200 240 300 500

Using the coefficients from Table D.11:

$$\begin{aligned} b &= 0.5739(500-110) + 0.3291(300-120) + 0.2141(240-130) \\ &\quad + 0.1224(200-140) + 0.0399(160-150) \\ &= 223.82 + 59.24 + 23.55 + 7.34 + 0.40 \\ &= 314.35 \end{aligned}$$

Since s = 119.47 for these data,

$$Calc\ W = \frac{314.35^2}{(10-1)119.47^2} = 0.769$$

Table D.12 gives *Tab W* = 0.842. Since *Calc W* < *Tab W*, reject normality (as we thought we should do in section 11.2).

Example (n *odd this time*)

Test the normality of the following sample of 9 observations (data from section 9.4):

163 157 160 168 155 168 164 157 169

In rank order these are:

155 157 157 160 163 164 168 168 169

The coeffients in Table D.11 give:

$$\begin{aligned} b &= 0.5888(169-155) + 0.3244(168-157) + 0.1976(168-157) \\ &\quad + 0.0947(164-160) \end{aligned}$$

$= 8.24 + 3.57 + 2.17 + 0.38$
$= 14.36$

Since $s = 5.34$ for these data,

$$Calc\ W = \frac{14.36^2}{(9 - 1)5.34^2} = 0.904$$

Tab W = 0.829, and hence *Calc W* > *Tab W*, so we do not reject normality. We conclude that these data are consistent with a normal distribution (as we thought we should do in section 9.4).

15.6 SUMMARY

Goodness-of-fit tests are tests to decide whether it is reasonable to conclude that a sample of univariate (one-variable) data could have been drawn from a particular type of distribution. Four types of distribution were covered, namely the simple proportion, binomial, Poisson and the normal. The first three were tested using a χ^2 test, the last by means of the Shapiro–Wilk test.

For the χ^2 test, the sample data are in the form of observed frequencies. Expected frequencies are calculated assuming that the sample data do come from the particular distribution under investigation. The degrees of freedom for *Tab* χ^2 are, in general, equal to (number of categories after combinations) − (number of distribution parameters estimated from the sample data) − 1.

WORKSHEET 15: GOODNESS-OF-FIT TESTS

1. In Mendel's experiments with peas, he classified 556 peas into four categories as follows:

Type of pea	*Number of peas*
Round and yellow	315
Round and green	108
Wrinkled and yellow	101
Wrinkled and green	32

   Are these data consistent with Mendel's theory of heredity that these categories should occur in the proportions 9:3:3:1?

2. The number of fatal road accidents in one year in a large city were tabulated according to the time they occurred:

Time	00.00–04.00	04.00–08.00	08.00–12.00	12.00–16.00	16.00–20.00	20.00–24.00
Number of accidents	28	15	14	18	15	30

Test the hypotheses that:

(a) Accidents are uniformly distributed in time.

(b) Accidents occur in the ratios 2:1:5:4:5:3, these being the estimated ratios of the volumes of traffic occurring in the city for the six four-hour periods.

3. The number of sheep farms of a given size in a county and the type of land on which they were situated were as follows:

Type of land	*Number of sheep farms*
Flat	43
Hilly	32
Mountainous	5

If 35% of the county is flat, 50% is hilly and 15% is mountainous, is the number of farms independent of the type of land?

4. A random sample of 100 families were asked how many cars they owned. The results were:

Number of cars	0	1	2 or more
Number of families	35	45	20

Test the hypothesis that, for all families, the ratios are 1:2:1 for the three categories of the number of cars owned.

5. For a random sample of 300 families each with three children, the distribution of the number of boys was as follows:

Number of boys	0	1	2	3
Number of families	55	108	102	35

Test the hypothesis that the number of boys in families with three children has:

(a) A binomial distribution with $p = 0.5$, implying that boys and girls are equally likely at each birth.

(b) A binomial distribution. *Hint*: estimate p from the sample data using the relative frequency definition, that is:

$$\frac{\text{total number of boys in the 300 families}}{\text{total number of children in the 300 families}}$$

Compare the conclusions of (a) and (b).

6. Samples of 10 pebbles were taken at random from each of 200 randomly selected sites on a beach. The number of limestone pebbles in each sample was counted. The results are summarized in the following frequency distribution table:

Number of limestone pebbles	0	1	2	3	4	5	6	7	8	9	10
Number of sites	0	7	20	45	53	39	25	8	3	0	0

How would you have selected the sites randomly? How would you select a sample of 10 pebbles from a particular site? Is it reasonable to conclude that the number of limestone pebbles in samples of 10 has a binomial distribution with $p = 0.4$?

7. An experiment was carried out to test whether the digit '8' occurred randomly in tables of random numbers. Successive sets of 20 single digits (0, 1, 2, . . . , 9) were examined and the number of times the digit '8' occurred was noted for each set.

Number of '8' digits found	0	1	2	3	4	5	6 or more
Number of sets	25	45	70	35	15	10	0

What conclusions can be drawn from these data?

8. A survey was conducted to decide whether a particular plant species was randomly distributed in a meadow. Eighty points in a meadow were randomly selected. A quadrat was placed with its centre at each selected point and the number of individual plants of the species was noted:

Number of individual plants per quadrat	0	1	2	3	4	5	6
Number of quadrats	11	37	12	7	6	4	3

How would you have selected 80 points randomly in a meadow? Is it reasonable to conclude that the plant species was randomly distributed in the meadow?

9. The number of dust particles occurring in unit volumes of a gas was counted. The procedure was repeated 100 times for the same constant volume. Given the following results, is it reasonable to assume

that the number of dust particles per unit volume is randomly distributed with a mean of two particles per unit volume?

Number of particles	0	1	2	3	4 or more
Number of times this number of particles observed	9	32	26	15	18

10. The number of minor defects noted by an inspector in 90 cars leaving a production asssembly line was as follows

Number of defects	0	1	2	3	4	5	6
Number of cars	35	13	6	5	18	10	3

(a) Test first whether the mean and variance of the number of defects are approximately equal. (This is a quick but not very reliable test for a Poisson distribution.)

(b) Now use the χ^2 test to decide whether the number of defects per car is randomly distributed.

11. Test the following data sets for normality:

(a) The ten differences listed in Table 9.1.

(b) The A-level counts of the nine BA students shown in Fig. 9.6, also listed in column 7 of Appendix A.

Appendix A

Data set for a random sample of 40 students

Student reference number	*Sex 1 = Male 2 = Female*	*Height (cm)*	*Number of Siblings*	*Distance from home to Oxford (km)*	*Type of degree 1 = BA 2 = BSc*	*A-level count*
1	1	183	1	80	2	6
2	2	163	2	3	1	32
3	2	152	2	90	1	22
4	2	157	3	272	2	12
5	2	157	1	80	2	12
6	2	165	3	8	2	18
7	1	173	1	485	2	14
8	1	180	2	176	2	8
9	2	164	2	10	2	6
10	2	160	3	72	1	18
11	2	166	0	294	2	16
12	2	157	1	22	1	12
13	2	168	0	144	2	12
14	2	167	2	160	2	12
15	2	156	1	50	2	10
16	2	155	1	64	1	12
17	1	178	1	224	2	8
18	2	169	3	480	2	10
19	2	171	5	56	2	6
20	1	175	3	141	2	8
21	1	169	2	259	2	4
22	2	168	4	96	2	6
23	2	165	1	104	2	12

Student reference number	*Sex 1 = Male 2 = Female*	*Height (cm)*	*Number of Siblings*	*Distance from home to Oxford (km)*	*Type of degree 1 = BA 2 = BSc*	*A-level count*
24	2	166	1	90	2	8
25	2	164	3	72	2	22
26	2	163	1	37	2	8
27	2	161	1	208	2	6
28	2	157	2	40	2	18
29	1	181	2	120	2	10
30	2	163	1	400	2	10
31	2	157	2	208	1	10
32	2	169	2	169	2	12
33	2	177	2	410	1	16
34	2	174	1	90	2	10
35	1	183	1	80	2	10
36	1	181	2	278	2	8
37	1	182	1	240	2	6
38	1	171	9	192	2	24
39	1	184	2	35	1	14
40	1	179	1	45	1	10

Appendix B

Multiple-choice test

Answer the following 50 questions by writing a, b or c according to your choice for each question; if you do not know the answer write d. You may use a calculator and statistical tables. Allow up to $1\frac{1}{2}$ hours.

If you want to see how well you scored, give yourself 2 marks for a correct answer, −1 mark for an incorrect answer and 0 marks for a 'don't know'. (The marking means that the maximum mark is 100, and the average mark for 'guessers' is 0.)

1. Which of the following is a continuous variable?
   (a) The time between stimulus and reaction.
   (b) The colour of a person's hair.
   (c) The number of errors on one page of *The Guardian*.

2. Which diagram is used to represent continuous data?
   (a) A histogram.
   (b) A bar chart.
   (c) A line chart.

3. The lower quartile of a distribution is such that:
   (a) $\frac{1}{4}$ of the values are greater than it.
   (b) $\frac{1}{4}$ of the values are less than it.
   (c) $\frac{3}{4}$ of the values are less than it.

4. The average which represents the value of a total when shared out equally is the:
   (a) mean.
   (b) median.
   (c) mode.

5. Half the values in a set of data are less than the:
   (a) mean.
   (b) median.
   (c) lower quartile.

6. For a symmetrical distribution:
   (a) mode = median = mean.
   (b) mode > median > mean.
   (c) mode < median < mean.

7. Which summary statistics are preferred when the distribution is roughly symmetrical?
   (a) Median and inter-quartile range.
   (b) Mode and range.
   (c) Mean and standard deviation.

8. $\frac{\Sigma x^2}{n}$ means:
   (a) Sum the n observations of x, square, and divide by n.
   (b) Sum the n observations of x, divide by n, and square.
   (c) Square each of the n observations of x, sum, and divide by n.

9. Three cards are drawn without replacement from a well shuffled pack. The probability that they are all diamonds is:
   (a) 1/64.
   (b) 33/2704.
   (c) 11/850.

10. A box contains 10 balls, 5 red and 5 white. The probability that two white balls are drawn with replacement is:
    (a) 1.
    (b) 1/4.
    (c) 1/5.

11. A box contains 2 red, 2 black and 2 green balls. The probability that two balls of the the same colour are drawn without replacement is:
    (a) 1/3.
    (b) 1/5.
    (c) 1/7.

12. $P(A|B)$ means:
    (a) The probability of A divided by B.
    (b) The probability of A conditional on B.
    (c) The probability of B conditional on A.

13. Two events A and B are such that, if B occurs the probability of A is unchanged. The events are said to be:
    (a) mutually exclusive.
    (b) exhaustive.
    (c) statistically independent.

14. Two independent events A and B have probabilities $P(A) = 1/3$, $P(B) = 1/4$. P(A or B or both) is:
    (a) 7/12.
    (b) 2/7.
    (c) 1/2.

15. A and B are mutually exclusive events, $P(A) = 1/4$ and $P(B) = 1/3$. P(A or B) is:
    (a) 7/12.
    (b) 1/2.
    (c) 1/12.

16. Two events A and B are independent if:
    (a) $P(A) = P(B)$.
    (b) $P(A|B) = P(A)$.
    (c) P(A or B) = $P(A) + P(B)$.

17. Random numbers are uniformly distributed over the range 0 to 1. The probability of getting a random number between 0.4 and 0.7 is:
    (a) 0.4.
    (b) 0.7.
    (c) 0.3.

18. In order to test the effectiveness of a drug, 12 individuals are treated with it. The variable of interest is the number of people who recover after taking the drug. The appropriate distribution to use in this case is the:
    (a) binomial.
    (b) normal.
    (c) Poisson.

19. The binomial distribution is:
    (a) always symmetrical.
    (b) sometimes symmetrical.
    (c) never symmetrical.

20. For a binomial distribution with $n = 20$, $p = 0.25$, the probability of 3 or fewer successes is:
    (a) 0.2252.
    (b) 0.9087.
    (c) 0.0913.

21. For a binomial distribution with $n = 5$, $p = 0.1$, the probability of exactly 3 successes is:
    (a) 0.0081.
    (b) 0.6.
    (c) 0.5.

22. The distribution of the number of random events per unit time is the:
    (a) Poisson.
    (b) normal.
    (c) binomial.

23. For a Poisson distribution with a mean $m = 10$ per unit time, the probability of at least 23 random events per unit time is:
    (a) 0.0003.
    (b) 0.0002.
    (c) 0.0004.

24. The distribution of the heights of adult male humans is likely to be:
    (a) binomial.
    (b) Poisson.
    (c) normal.

25. In a normal distribution with mean μ and standard deviation σ:
    (a) 10% of the values are outside the range $(\mu - 1.645\sigma)$ to $(\mu + 1.645\sigma)$.
    (b) 10% of the values are greater than $(\mu + 1.645\sigma)$.
    (c) 10% of the values are outside the range $(\mu - 1.96\sigma)$ to $(\mu + 1.96\sigma)$.

26. In a normal distribution with $\mu = 10$, $\sigma^2 = 4$, the probability of exceeding 13 is:
    (a) 0.0668.
    (b) 0.2266.
    (c) 0.9332.

27. A random sample is one in which:
    (a) some members of the population are more likely to be picked than others.
    (b) each member of the population has an equal chance of being picked.
    (c) members of the population are picked because they are thought to be representative.

28. A random sample is preferred by statisticians because it:
    (a) is representative of the population.
    (b) ensures against bias.
    (c) is chosen in a special statistical way.

29. The distribution of the means of samples of size n, taken from a population with a standard deviation of σ, has a standard deviation of:
    (a) σ/n.
    (b) $\sigma/\sqrt{n}$.
    (c) σ.

30. A 90% confidence interval for the mean of a population is such that:
    (a) 10% of the values in the population lie outside it.
    (b) there is a 90% chance that it contains all the values in the population.
    (c) there is a 90% chance that it contains the mean of the population.

31. If the sample size is increased the 95% confidence interval for the population mean will:
    (a) increase.
    (b) decrease.
    (c) remain the same.

32. The critical value of U for a Mann–Whitney test with samples of 7 and 8 for a two-tailed test at the 5% level is:
    (a) 6.
    (b) 5.
    (c) 10.

33. A one-tailed test is used if the alternative hypothesis is:
    (a) H_1: $\mu_1 = \mu_2$.
    (b) H_1: $\mu_1 \neq \mu_2$.
    (c) H_1: $\mu_1 > \mu_2$.

34. If we decide not to reject a null hypothesis H_0 this:
    (a) provcs that H_0 is true.
    (b) proves that H_1 is false.
    (c) implies that H_0 is likely to be true.

35. The critical value of T for a Wilcoxon signed rank test with 20 pairs for a one-tailed test at the 1% level of significance is:
    (a) 37.
    (b) 43.
    (c) 238.

36. The t test for samples from a normal population must be used when:
    (a) the sample size is small.
    (b) the standard deviation is unknown.
    (c) the sample size is small and the standard deviation is unknown.

37. If the calculated value of χ^2 is less than the tabulated value of χ^2:
    (a) H_0 is rejected.
    (b) H_0 is not rejected.
    (c) H_1 is accepted.

38. If the calculated value of Wilcoxon T is less than or equal to the tabulated value of T, you should:
    (a) reject H_0.
    (b) not reject H_0.
    (c) accept H_1.

39. In analysing the results of an experiment in which individuals in two samples, both of size 6, have been paired, the calculated value of t will have:
    (a) 11 degrees of freedom.
    (b) 5 degrees of freedom.
    (c) 10 degrees of freedom.

40. The Wilcoxon signed rank test is preferred to the t test when:
    (a) the sample sizes are large.
    (b) the data are paired.
    (c) the assumptions of the t test do not hold.

41. The unpaired samples t test is prefered to the Mann–Whitney U test if:
    (a) the sample sizes are small.
    (b) the sample sizes are equal.
    (c) the assumptions of the t test are valid.

42. For a contingency table with 4 rows and 4 columns, the tabulated value of χ^2 at the 5% level of significance is:
    (a) 16.9.
    (b) 19.0.
    (c) 7.82.

43. The expected frequencies used in a χ^2 test on data in a contingency table must be:
    (a) whole numbers.
    (b) all greater than or equal to 5.
    (c) greater than the corresponding observed frequencies.

44. A correlation coefficient of 0.80 between two variables implies:
    (a) that as one variable increases the other decreases.
    (b) that H_0: $\rho = 0$ should be rejected in favour of H_1: $\rho > 0$.
    (c) nothing, since there is insufficient information.

45. A significantly high positive value of a correlation coefficient between two varables implies:
    (a) that as one variable increases, the other variable decreases.
    (b) a definite causal relationship.
    (c) a possible causal relationship.

46. A random sample of 12 pairs of values from two normally distributed populations has a correlation coefficient of 0.54. Hence there is:
    (a) a significant correlation between the two variables.
    (b) no significant correlation between the two variables.
    (c) no significant correlation between the two variables at the 5% level.

47. If x is measured in centimetres and y is measured in grams, the correlation coefficient between x and y will:
    (a) have no units.
    (b) be in cm $\times$ g units.
    (c) be in cm or g units.

48. Linear regression is used to:
    (a) prove that there is a causal relationship between two variables.
    (b) predict one variable from the other.
    (c) show that as one variable increases so does the other.

49. The slope (or gradient) of a regression line:
    (a) always lies between -1 and $+1$.
    (b) can never be negative.
    (c) can have any value.

50. The purpose of calculating Pearson's r is to:
    (a) replace points on a scatter diagram by a straight line.
    (b) measure the degree to which two variables are linearly associated.
    (c) predict one variable from another variable.

Appendix C

Solutions to worksheets and multiple-choice test

WORKSHEET 1 (Solutions)

Key: C = continuous, D = discrete, R = ranked, Cat = categorical.

1. C; 0–1000 hours; lightbulb.
2. D; 0,1, 2, . . . , 1000; year.
3. C; £20–£50; hotel.
4. D; 0, 1, 2, . . . , 100; hotel.
5. Cat; professional, skilled, . . . ; adult male.
6. D; 0, 1, . . . , 10; 100 hours.
7. C; 0–100 hours; 100 hours.
8. D; 0, 1, 2, . . . , 1000; month.
9. R; 1, 2, 3, . . . , 30; annual contest.
10. C; 0–300 cm; county.
11. D; 0, 1, . . . , 20; year (in period 1900–1993).
12. D; 0, 1, . . . , 10^6; oil rig (output in barrels).
13. D; 0, 1, . . . , 1000; 10-second period.
14. D; 0, 1, . . . , 10; series of 10 encounters with T-junction.

15. R; A, B, C, D, E, N, U; candidate.
16. Cat; black, brown, . . . ; person.
17. Cat; presence, absence; square metre of meadow.
18. C; 0–5 seconds; rat.
19. D; 0, 1, 2, . . . , 20; page.
20. C; 0–10 kg; tomato plant.
21. Cat; sand, limestone, etc.; core sample.
22. C; 0–100%; sample.
23. Cat; Con., Lab., Lib-Dem., . . . ; person.

WORKSHEET 2 (Solutions)

1. (a) −1.8
   (b) 3
   (c) 33.58
   (d) 0.000 335
   (e) 0.1575
   (f) 341.297
   (g) 8.35
   (h) 5.29
   (i) 0.125
   (j) 0.0164
   (k) 1
   (l) 125
   (m) 4.953
   (n) 0.2019
   (o) 0.839
   (p) 8
   (q) 24, 1, 720, 1, not defined, not defined.
2. (a) 33.6
   (b) 0.000 33
   (c) 0.16
   (d) 341.3
   (e) 300
3. (a) 55
   (b) 3
   (c) 55
   (d) 44
4. 24, 3, 576, 84, 0, 12, 12.
5. 17.7, 3.54, 313.29, 85.27, 0, 22.612, 22.612.

WORKSHEET 3 (Solutions)

1. (a) Group frequency table plus histogram, or cumulative frequency table plus cumulative frequency polygon.
   (b) Group frequency table plus bar chart.
   (c) Simple list plus bar chart.
   (d) As for (a).
   (e) As for (a).
   (f) Group frequency table with 11 groups, line chart.
   (g) As for (a).
   (h) Group frequency table with 6 groups, line chart.
   (i) Group frequency table with 6 groups, bar chart.
   (j) Simple list plus dotplot.
   (k) Group frequency table with 3 groups, bar chart.
   (l) As for (a).

2. (a) The dotplot ranges from 0.2 to 5.7 with a mean of about 1.5 g. It is reasonably symmetrical, except for two outliers at 4.6 and 5.7, and it is reasonably bunched in the middle.

Weight (g)	*No. of jars*
0.00–0.99	13
1.00–1.99	25
2.00–2.99	20
3.00–3.99	9
4.00–4.99	2
5.00–5.99	1

The histogram also shows some positive skewness.

```
MTB> SET  C1
DATA> 0.7 1.3 . . . 4.6
DATA> 1.9 1.7 . . . 2.8
:
.
DATA> 3.0 1.6 . . . 2.7
DATA> END
MTB> DOTPLOT  C1
MTB> STEM-AND-LEAF  C1
MTB> HISTOGRAM  C1;
SUBC> INCREMENT  1;
SUBC> START  0.495.
```

(b) Add the following commands to those in (a):

```
MTB> SET  C2
DATA> 1 1 1 1 1 1 1 1 1 1
DATA> 1 1 1 1 1 1 1 1 1 1
DATA> 1 1 1 1 1 1 1 1 1 1
```

```
DATA> 1 1 1 1 1 2 2 2 2 2
DATA> 2 2 2 2 2 2 2 2 2 2
DATA> 2 2 2 2 2 2 2 2 2 2
DATA> 2 2 2 2 2 2 2 2 2 2
DATA> END
MTB> DOTPLOT  C1;
SUBC> BY  C2.
```

There is not much difference between the two dotplots. The second shift's values are less variable, but only because of the two outliers in the data from the first shift.

3.

Distance (km)	*No. of students*
0– 49.9	8
50– 99.9	12
100–149.9	4
150–199.9	4
200–249.9	4
250–299.9	4
300–349.9	0
350–399.9	0
400–449.9	2
450–499.9	2

Distance (km)	*Cumulative no. of students*
less than 0	0
50	8
100	20
150	24
200	28
250	32
300	36
350	36
400	36
450	38
500	40

$1/2 \times 40 = 20$, and 20 students live less than 100 km from Oxford, according to the above table. From the raw data, the 20th values is 96, and the 21st value is 104, so again the median is 100 km.

4. There are 31 BSc and 9 BA students. Using dotplots, we can say that the A-level count is generally lower for BSc students. (Approx. means are 11 and 16, although one BA outlier means that we should be comparing medians instead.

Approx. medians are 10 and 13 respectively.) For a more sophisticated analysis of these data, see sections 9.11 and 10.13 in this book.

WORKSHEET 4 (Solutions)

1. (a) They measure a middle value for a set of data.
   (b) Sample mode.
   (c) Sample mean.
   (d) Sample median.
   (e) Markedly skew data.
   (f) Only for categorical data (for which the mean is not defined).
   (g) Roughly symmetrical data.

2. (a) Because averages can be misleading, and to show how much data vary.
   (b) (i) Sample standard deviation, (ii) Sample inter-quartile range.
   (c) Question 3 below.
   (d) (i) Sample standard deviation, (ii) Sample inter-quartile range, (iii) None.
   (e) (i) Sample mean, (ii) Sample median, (iii) None.

3. (a) $\bar{x} = 78.1$, sample median $= 67$, sample mode $= 62$ or 67. Answers differ because of one extremely high value, so positive skewness, and hence mean > median. Mode is non-unique.
   (b)

$$s = \sqrt{\frac{89\,797 - \dfrac{937^2}{12}}{11}} = 38.9.$$

   Sample inter-quartile range $= 73.25 - 62 = 11.25$.
   Answers differ because these two measures (standard deviation and inter-quartile range) measure variation in different ways.
   (c) Sample median and sample inter-quartile range, because of skewness in data. Coefficient of skewness $= 3(78.1 - 67)/38.9 = 0.856$.
   Alternatively, summarize by stating that one value was £200 and the other eleven are summarized by $\bar{x} = 67.0$, $s = 6.47$.

4.

$$\bar{x} = \frac{11.44}{11} = 1.04, \quad s = \sqrt{\frac{11.9068 - \dfrac{11.44^2}{11}}{10}} = 0.0303.$$

5. (a) $\bar{x} = 1.9357$, $s = 1.0699$.
   (b) $(n + 1)/2 = 35\frac{1}{2}$, so median $= 1.8$ (35th and 36th value are both 1.8).
   $(n + 1)/4 = 17\frac{3}{4}$, so Q_1, the lower quartile $= 1.275$ (17th value is 1.2, 18th is 1.3).
   $3(n + 1)/4 = 52\frac{1}{4}$, so Q_3, the upper quartile $= 2.7$ (52nd and 53rd value are both 2.7).
   Hence inter-quartile range $= Q_3 - Q_1 = 2.7 - 1.275 = 1.425$.

A measure of skewness $= 3(1.9357 - 1.8)/1.0699 = 0.38$, so not very skew. Preferred average is the mean, preferred measure of variation is the standard deviation.

6. (a) $\bar{x} = 151.9$, $s = 127.6$
   (b) median = 100, inter-quartile range = 220 − 58 = 162. A measure of skewness = 1.22, which indicates marked positive skewness, hence preferred average is the median and the preferred measure of variation is the inter-quartile range.

7. Put the distance data into C1, and the sex data into C2 using the command READ C1 C2. Then type in:

```
MTB> NAME  C1  'DIST'
MTB> NAME  C2  'SEX'
MTB> DESCRIBE  C1;
SUBC> BY  C2.
MTB> DOTPLOT  C1;
SUBC> BY  C2.
MTB> BOXPLOT  C1;
SUBC> BY  C2.
```

WORKSHEET 5 (Solutions)

1. Refer to section 5.3 and section 5.4.

2. Using P(E') = 1 − P(E), where E is success, E' is 'not success', i.e. failure, P (failure) = 1 − 0.2 = 0.8.

3. The three events '2 heads', '1 head and 1 tail', '2 tails' are not equally likely. However, the events *HH*, *HT*, *TH* and *TT* are equally likely and each has a probability of 1/4 of occurring. P(2 heads) = P(*HH*) = 1/4, P(1 head) = P (*HT* or *TH*) = P(*HT*) + P(*TH*) = 1/4 + 1/4 = 1/2 (the events *HT* and *TH* are mutually exclusive), P(2 tails) = P(*TT*) = 1/4

4. Our estimate of the probability of heads is 1 after 5 tosses, but the number of trials is not large enough for an accurate estimate to be made.

5. (a) Each die has the same probability (= 1/3) of being selected.
   (b) *Y*1, *Y*2, . . . , *Y*6, *B*1, *B*2, . . . , *B*6, *G*1, *G*2, . . . , *G*6; 18 outcomes in all.
   (c) Yes.
   (d) 1/18.
   (e) (i) 6/18, (ii) 3/18, (iii) 9/18, (iv) 3/18. (v) 12/18.

6. (a) 9/27, (b) 1, where 9/27 = Area of rectangle with base 164.5 to 169.5/Total area of histogram.

7. Suppose that the number of *U*'s after 1, 2, 3, 4, 5, 10, 20, 30, 40, 50 tosses are as follows:

No. of tosses	1	2	3	4	5	10	20	30	40	50
No. of *U*'s	1	1	2	3	3	5	11	17	23	29
Est. of P(*U*)	1	0.5	0.67	0.75	0.60	0.50	0.55	0.57	0.58	0.58

8. (a) A', (b) $A|B$, (c) $B|A$.

9. (a) Probability of event *A* given that event *B* has occurred.
   (b) Probability of event *B* given that event *A* has occurred.

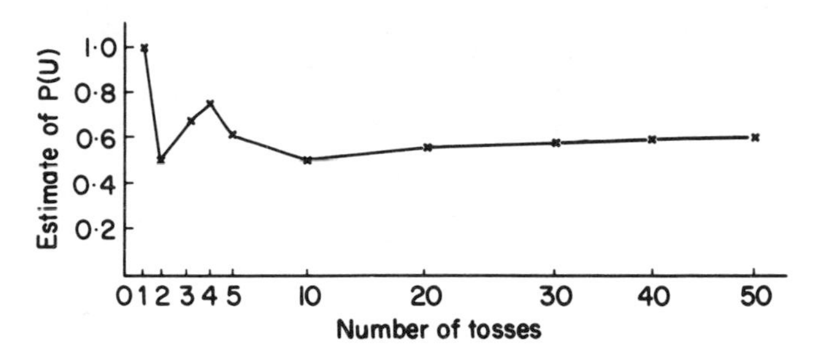

Fig. C.1 Estimating the probability that a drawing pin will fall point upwards.

(c) Probability that event A will not occur.
(d) The probability of event A is not affected by whether event B has occurred, and vice versa.
(e) If event A occurs, event B cannot, and vice versa.

10. $P(A \text{ and } B) = P(A)P(B|A)$, or $P(A \text{ and } B) = P(B)P(A|B)$.
If A and B are statistically independent, $P(A \text{ and } B) = P(A)P(B)$.

11. $P(A \text{ or } B \text{ or both}) = P(A) + P(B) - P(A \text{ and } B)$.
If A and B are mutually exclusive, $P(A \text{ or } B) = P(A) + P(B)$.

12. (a) A and B are statistically independent.
(b) A and B are mutually exclusive.

13. P(3 or 6) = P(3) + P(6) = 1/6 + 1/6 = 1/3.

14. P(red or picture or both) = P(red) + P(picture) − P(red and picture)
= 26/52 + 16/52 − 8/52
= 34/52.

15. The probability tree has 8 branch-ends, each having a probability of 1/8.
(a) 1/8, (b) 3/8, (c) 3/8, (d) 1/8.

16. P(at least one hit) = 1 − P(all miss) = 1 − 1/2 × 2/3 × 3/4 = 3/4.

17. $P(\text{exactly one defective}) = P(DD'D'D' \text{ or } D'DD'D' \text{ or } D'D'DD' \text{ or } D'D'D'D)$
$= P(DD'D'D' + P(D'DD'D') + P(D'D'DD') + P(D'D'D'D)$
$= 4(0.03 \times 0.97 \times 0.97 \times 0.97)$
$= 0.1095.$

18. From the information given, using C = car, H = house, use the laws of probability with $P(H) = 0.4$, $P(C) = 0.7$, $P(H \text{ and } C) = 0.3$.
(a) $P(H \text{ or } C \text{ or both}) = P(H) + P(C) - P(H \text{ and } C) = 0.4 + 0.7 - 0.3 = 0.8$.
(b) Since $P(H \text{ and } C) = P(C) \times P(H|C)$, $0.3 = 0.7 \times P(H|C)$.

Hence $P(H|C) = 0.3/0.7 = 0.43$.

19.

	Bedroom type		
	S	*D*	
Bath	2	9	11
No bath	4	5	9
	6	14	20

(a) 11/20 = 0.55.
(b) 2/11 = 0.18.

20. First member seems to be using $P(U_1 \text{ and } U_2) = P(U_1)P(U_2) = 1/4 \times 1/4$, where U_1, U_2 are the events 'unoccupied on first visit', 'unoccupied on second visit', respectively.
Therefore P(occupied on either 1st or 2nd visit) $= 1 - (1/4)^2 = 15/16$. This argument is incorrect because it assumes independence, while it makes more sense to assume that if a house is unoccupied on one visit, the chances of it being unoccupied on a second visit will be affected.
Second member seems to be adding the 1/4's. Is he using $P(U_1 \text{ or } U_2) = P(U_1) + P(U_2)$? If so, he is incorrect because U_1 and U_2 are not mutually exclusive.
The correct argument is $P(U_1 \text{ and } U_2) = P(U_1)\, P(U_2|U_1) = 1/4\, P(U_2|U_1)$. We don't know $P(U_2|U_1)$ but it must be between 0 (if unoccupied houses on the first visit are all different from those unoccupied on the 2nd visit) and 1 (if the unoccupied houses on the first visit are all exactly the same as those unoccupied on the second visit). Hence required probability lies between $1 - 1/4 \times 0$ and $1 - 1/4 \times 1$, i.e. between 1 and 3/4.

21. (a) P(success) = P(lift-off success) × P(separation success | lift-off success) × P(mission completed | separation success)
$= (1 - 0.1)(1 - 0.05)(1 - 0.03)$
$= 0.8294$.
(b) P(failure) $= 1 - 0.8294 = 0.1706$.

22. (a) 13/40
(b) 27/40
(c) 26/40
(d) (i) 13/13, (ii) 12/27
(e) (i) 13/25, (ii) 0/15, no, since P(male) = 13/40, while P(Male $>$ 165) = 13/25, and these are not equal.
(f) 11/40
(g) P(male or BSc or both) = P(male) + P(BSc) − P(male and BSc)
= 13/40 + 31/40 − 11/40
= 33/40.

WORKSHEET 6 (Solutions)

1. n is the number of trials, p is the probability of success in a single trial.
2. By checking the four conditions (section 6.3).
3. The choice is arbitrary, although to use Table D.1 it is better to choose the outcome with a probability of less than 0.5 as the 'success'.
4. The number of successes in n trials.
5. (a) 2, (b) 3, (c) 4, (d) 1.5, 0.866, (e) yes, because $p = 0.5$.
6. 0.0625, 0.2500, 0.375, 0.2500, 0.0625. Because the five events are mutually exclusive and exhaustive. Expect 12 (or 13), 50, 75, 50, 13 (or 12) families.
7. $B(20, 0.2)$, (a) 4, (b) 0.0321, (c) 0.9885.
8. (a) 0.9231, (b) 0.2611, (c) 0.0002.
9. 0.0313, 0.1563, 0.3125, 0.3125, 0.1563, 0.0313 (using Table D.1).
10. 0.3828, probability.
11. (a) Because the four conditions are satisfied.
    (b) $n = 20$, $p = 0.2$.
    (c) (i) 0.0115, (ii) 0.9885.
12. $B(6, 0.1)$, (a) 0.5314, (b) 0.4686, where trial is 'examining an egg'. A 'success' is 'cracked egg'.
    $B(5, 0.5314)$ where a trial is now 'examining a box', and 'success' is 'a box with no cracked eggs'. P(3 or more) = P(3) + P(4) + P(5) = 0.5587.
13. m is the mean number of random events per unit time or space.
    e is a constant, the number 2.718 . . .
    x is the number of random events per unit time or space.
    x can take values, 0, 1, 2, . . .
14. Standard deviation = 2, variance = 4. Mean and variance are equal in value.
15. In each unit of time (or space) the probability that the event will occur is the same.
16. $m = 1/10 = 0.1$ breakdowns per week. 0.9048, 0.0905, 0.0045.
17. $m = 1/5 = 0.2$ misprints per page. P(0) = 0.8187 or 81.87%. Expect no misprints on 409 pages, since $0.8187 \times 500 = 409$.
18. $m = 2.5$, (a) 8.21%, (b) 45.62%, (c) P(4 or more) = 0.2424 or 24.24%.
19. $m = 3$. (a) 0.0498, (b) 0.2240, (c) 0.5768.
20. 0.634, 0.6321. Good agreement, first is correct, second is only an approximation, but an excellent one.
21. $m = np = 4$, P(0) = 0.0183.

22. (a) 0.0245, 0.0318, not such good agreement. Answers using Poisson approximation are less accurate for small probabilities.
    (b) 0.1849, 0.1755, good agreement.

WORKSHEET 7 (Solutions)

1. Mean, standard deviation.
2. No, always symmetrical.
3. The total area under the curve for the normal distribution is 1, the total area of the rectangle for the rectangular distribution is also 1.
4. (a) 4.75%, (b) 0.05%, (c) 95.2%. Expect 2, 0, 48 (to nearest whole orange). With $\mu = 70 - 5 = 65$, $\sigma = 3$, new answers are: (a) 0.05%, (b) 4.75%, (c) 95.2%. Expect 0, 2, 48.
5. For the seven grades, percentages are 0.62%, 6.06%, 24.17%, 38.30%, 24.17%, 6.06%, 0.62%. Total price of 10 000 oranges = $(62 \times 4 + 606 \times 5 + \cdots + 62 \times 10) = 70\,000$p. Mean price = 7p.
6. Percentage rejected = 18.15%. With new mean of 0.395, percentage rejected = 13.36%. This is a minimum because 0.395 is halfway between the rejection values of 0.38 and 0.41.
7. (a) 203, (b) 19, (c) 778.
8. For left-handed area of $1 - 0.15 = 0.85$, $z = 1.04$, using Table D.3(a) in reverse. Hence $(x - \mu)/\sigma = (85 - 65)/\sigma = 1.04$, so $\sigma = (85 - 65)/1.04 = 19.2$ cm (Fig. C.2).
   For 50 cm, $z = -0.78$, so 21.8% of years have less than 50 cm, using Table D.3.
9. For a left-handed area of 0.2, $z = -0.84$. Hence $(x - \mu)/\sigma = (30 - \mu)/4 = -0.84$. It follows that $\mu = 30 + 4 \times 0.84 = 33.36$ (Fig. C.3). For £50, $z = 4.16$ and from Table D.3 (a) area to the left of $z = 4.16$ is virtually 1. So no staff earn more than £50 a week.
10. (a) 50%, (b) 95.25%, (c) 99.95%.

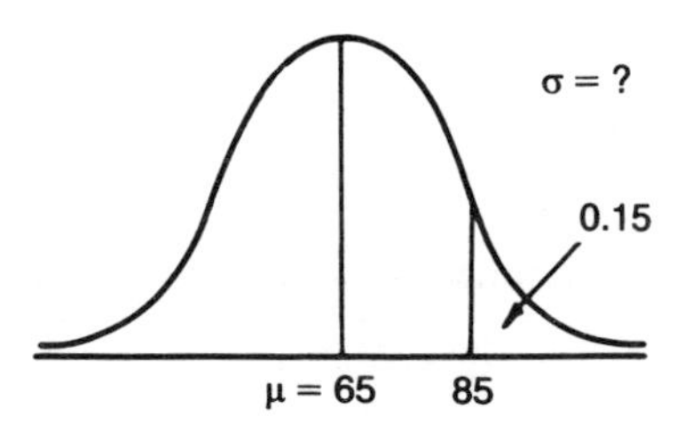

Fig. C.2 Distribution of the wages of a certain grade of kitchen staff.

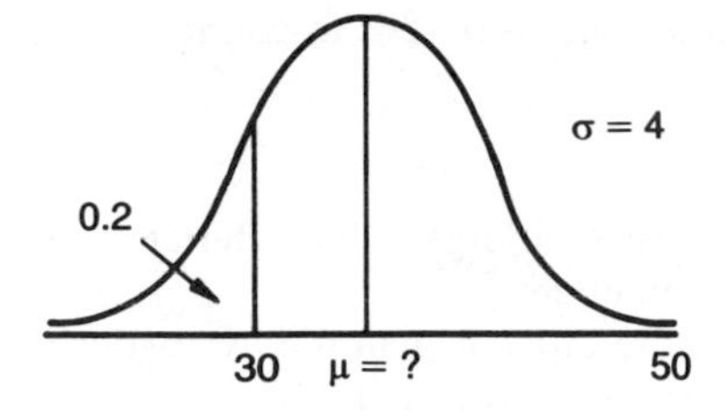

Fig. C.3 Distribution of the percentage of void space in sandstone.

11. $z = -2.33$, $(x - \mu)/\sigma = (x - 172)/8 = -2.33$, $x = 172 - 8 \times 2.33 =$ 153.4 cm.

12. (a) 50%, (b) 95.45%. New target mean is 25.82 kg, 0.1% exceed 27.37 kg ($z = 3.1$).

13. (a) $z = (x - \mu)/\sigma = ((\mu + \sigma) - \mu)/\sigma = 1$, and $z = ((\mu - \sigma) - \mu)/\sigma = -1$. Hence area = 0.8413 − (1 − 0.8413) = 0.6826.
    (b) Similar to (a) with $z = \pm 2$.
    (c) Similar to (a) with $z = \pm 3$.
    (d) Similar to (a) with $z = \pm 1.96$.
    (e) Similar to (a) with $z = +1.645$.

14. Use the normal approximation to the binomial with $n = 60$, $p = 0.8$, so $\mu = 48$, $\sigma = 3.1$.
    (a) P(>49.5) = 1 − 0.6844 = 0.3156.
    (b) P(49.5 − 50.5) = 0.1066.
    (c) P(<49.5) = 0.6844.

15. With a rectangular distribution, area to the right of 15 = 1/6, so late 1 day in 6. With a normal distribution, $z = 1.33$, area to the left of 15 = 0.9082, and hence the area to the right of 15 = 0.0918, so late 1 day in approximately 11 (Fig. C.4).

WORKSHEET 8 (Solutions)

1. (a), (b), (c), (d) – see sections 8.1, 8.2 and 8.3.
   (e) A census is a 100% sample, so the whole population is included in the

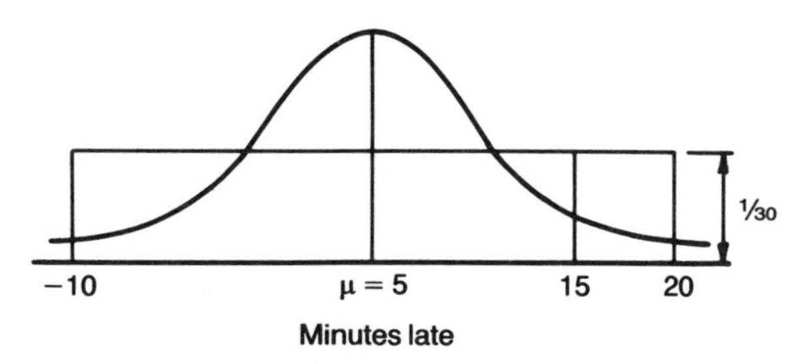

Fig. C.4 Two distributions of a commuter's lateness for work.

sample, and often the main purpose is to count the total number of individuals in the population.

2. See section 8.3.

3. (a) Might catch slowest and largest first. Better to number the mice 1 to 20 and use random numbers.
   (b) (i) Travellers more affluent than the average adult.
   (ii) Shoppers for food more representative of adults, might be biased in favour of housewives, etc.
   (iii) Adults leaving job centre may be unemployed, might be biased in connection with unemployment.
   (c) We do not know how the investigator actually 'randomly threw the quadrat'. Better to use grid method (Fig. 8.1), and place quadrat with its centre at the chosen points.
   (d) Use stratified sampling since there are three strata. Select 4, 5, 1 at random from three types of hotel, respectively.
   (e) Initial sample correctly chosen, but many of those chosen may not be on the telephone. Even if there is a telephone in the house, only one name appears in the telephone directory. Party worker should visit those selected (even then there could be problems – see Worksheet 5, Question 20).
   (f) People who visit a doctor's surgery are not typical of a random sample of his possible patients. Also those who volunteer may do so because they are prone to influenza, again not random. Better to choose a random sample from the alphabetical list of patients, assign half randomly to the vaccine and the other half to the placebo.

4.
```
MTB> RANDOM 108 C1 C2;
SUBC> INTEGER 1 6.
MTB> RMEAN C1 C2 INTO C3
MTB> PRINT C3
MTB> DESCRIBE C3
MTB> HISTOGRAM C3;
SUBC> START 1;
SUBC> INCREMENT 0.5.
```

WORKSHEET 9 (Solutions)

1. To give a measure of precision to a single value (or point) estimate of a population parameter, such as the mean (μ) or the proportion of successes (binomial p).

2. $\bar{x}$, s, n, and Table D.5. If n is small, variable must be approximately normally distributed.

3. False.

4. True.

5. True.

6. It does not imply '95% probability' (since either the population mean actually does lie between 10 and 12 and then the probability is 1, or it does not and then the probability is 0). Since, in repeated sampling, 95% of the 95% confidence intervals we calculate actually contain the mean, we feel that the confidence interval '10 to 12' has a very good chance of being one of those intervals which actually contains thc population mean. Think of betting on a horse at odds of 19 to 1 on. Alternatively, we can think of 'taking a risk of 5%' that the interval '10 to 12' does not contain the population mean.

7. £28 to £32. Need to sample 400 customers.

8. 72% to 88%. Need to ask 6150 people.

9. 125.3 to 132.5, (a) 7.2, (b) 9.5, (c) 3.6.

10. 9220 pebbles.

11. $\bar{x} = 308.3$, $s = 131.9$, 274.2 to 342.4.

12. 0.67 to 6.75 kg. The assumption that difference in weights is approximately normal is reasonable since weight is approximately normal. Also a dotplot of the data indicates approximate normality (symmetry plus concentration of points in the middle).

13. $s = 0.0982$, -0.37 to -0.58 for $(\mu_A - \mu_B)$. Assumptions: (i) Percentages are normally distributed, (ii) $\sigma_A = \sigma_B$, reasonable here since s_A and s_B are similar (see also solutions to Questions 16, 17, 18 of Worksheet 10).

14. $\bar{d} = -10.45$, $s_d = 10.13$, -3.6 to -17.3 for μ_d, where d = 'Town A − Town B' rainfall. Assumption that d is nomal, which looks reasonable from a dotplot.

15. $s = 4.00$, -4.0 to -14.3 for $(\mu_A - \mu_B)$. Assumptions: (i) Weights are normally distributed, reasonable if reasons for small variations are numerous and independent (see section 7.2). (ii) $\sigma_A = \sigma_B$. (Refer to solution to Question 13 above.)
    (a) 440.7 to 448.6.
    (b) 449.4 to 458.3.
    Reasonable for B, not for A, since 95% confidence interval for μ_B does contain 452, but 95% confidence interval for μ_A does not contain 452. (See also section 10.16, where this use of confidence intervals is discussed.)

WORKSHEET 10 (Solutions)

1. See sections 10.2 and 10.3.

2, 3, 4. See sections 10.4 and 10.14.

5. See section 10.3.

6. If we wish to decide whether the value of the parameter of interest is greater than (or less than) a particular value, then the alternative hypothesis is one-sided. If we wish to decide whether the value of the parameter is different

from (i.e. not equal to) a particular value, so the direction of the difference is not of interest, then the alternative hypothesis is two-sided.

7. *Calc* $t = 4.37$, *Tab* $t = 2.228$, data do not support stated hypothesis.
Assumption: Weight of sugar is 'approximately' normal. Reasonable if reasons for small variations in sugar weights are numerous and independent (see section 7.2), or we could draw a dotplot (see Fig. 3.1) to see if we have symmetry and concentration in the middle. The dotplot shows reasonable symmetry and an even spread, and hence is not severely non-normal.

8. *Calc* $t = 10.54$, *Tab* $t = 1.645$, manufacturer's claim is not justified.
Assumption: Approximate normality, but not important here because of large sample size.

9. $\bar{x} = 87.67$, $s = 10.81$, *Calc* $t = -1.18$, *Tab* $t = 1.699$, data do not support claim.
Assumption: Approximate normality of wages, which a histogram would indicate. In any case, the sample size is quite large.

10. *Calc* $z = 1.09$, *Tab* $z = 1.645$. Market share has not increased significantly.
Assumption: The four binomial assumptions, the main one being independence which means here that individuals do not influence each other in their choice of brand.

11. *Calc* $z = 3.46$, *Tab* $z= 1.96$. It is not reasonable to expect that 50% of all gourmets prefer thin soup.
Assumption: Similar to Question 10.

12. *Calc* $z = -2.55$, *Tab* $z = 1.645$. It is reasonable to conclude that the death rate is lower than 14%.
Assumption: The four binomial conditions, the main one being independence which means that chance of death is not influenced by others dying – probably true except for epidemics of fatal diseases (the Great Plague, for example).

13. (a) Growing conditions within a farm will be more homogeneous than between farms.
(b) $\bar{d} = 0.3143$, $s_d = 0.4140$, *Calc* $t = 2.01$, *Tab* $t = 2.447$. Mean yields not significantly different.
Assumption: Approximately normal differences; reasonable if reasons for small variations in differences are numerous and independent (see section 7.2). As in Question 7, a dotplot could be drawn, but would not be conclusive.

14. $\bar{d} = -0.66$, $s_d = 0.8591$, *Calc* $t = -1.72$, *Tab* $t = 2.132$. Allegation not supported by data.
Assumption: Approximately normal differences. No information on why variations occur, and very little data, so t test dodgy here.

15. $\bar{d} = 0.72$, $s = 0.7941$, *Calc* $t = 2.87$, *Tab* $t = 1.833$. Drug gives significantly more hours' sleep. Similar dotplot to that for Question 7.

16. $\bar{x}_1 = 4673.3$, $s_1 = 120.94$, $\bar{x}_2 = 4370.0$, $s_2 = 214.48$, $s = 174.1$, *Calc* $t = 3.02$, *Tab* $t = 2.228$. There is a significant difference in the mean strengths.

Assumptions: (a) Approximate normality in both populations, difficult to tell here and very little data. (b) $\sigma_1 = \sigma_2$. *Calc* $F = s_2^2/s_1^2 = 214.48^2/120.94^2 = 3.15$, *Tab* $F = 5.05$ (Table D.6 for 5, 5 d.f.) Since $3.15 < 5.05$, assumption of equal variances is reasonable (assuming normality, which is problematic as we have seen!).

17. $s = 0.5523$, *Calc* $t = 7.55$, *Tab* $t = 1.68$. It is reasonable to suppose that corner shops are charging significantly more on average than supermarkets.
    Assumptions: (a) Approximate normality, not important here because of large sample sizes (except as an assumption of the F test). (b) $\sigma_1 = \sigma_2$. *Calc* $F = 0.6^2/0.5^2 = 1.44$. *Tab* $F = 1.98$, for 24, 24 d.f. (Table D.6). Since $1.44 < 1.98$, assumption is reasonable. We would need dotplots of the raw data to check visually for normality.

18. $\bar{x}_A = 94.00$, $s_A = 28.17$, $n_A = 10$, $\bar{x}_B = 99.00$, $s_B = 27.47$, $n_B = 10$, $s = 27.82$, *Calc* $t = -0.40$, *Tab* $t = 1.73$. Mean amount of vanadium for area A is not significantly less than for area B.
    Assumptions: (a) Approximate normality in each population, difficult to tell here with small sample sizes, but dotplots do not indicate any extreme values and there is approximate symmetry. (b) $\sigma_A = \sigma_B$, *Calc* $F = 1.05$, *Tab* $F = 3.19$, so reasonable assumption.

19. (7) 1.02 to 1.06. Reject H_0: $\mu = 1$ since 1 outside the 95% confidence interval.
    (11) 54% to 66%. Reject H_0: p = 0.5, since 50% is outside the 95% confidence interval.
    (13) −0.07 to 0.70. Do not reject H_0: $\mu_d = 0$, since 0 is inside the 95% confidence interval.
    (16) 79.3 to 527.3. Reject H_0: $\mu_1 = \mu_2$, since 0 is outside the 95% confidence interval.

WORKSHEET 11 (Solutions)

1. Hypotheses, assumptions.
2. Assumptions, powerful.
3. Null, alternative, higher.
4. (a) Preference testing example.
   (b) Example where magnitudes of differences are known. Wilcoxon preferred.
5. Unpaired samples, powerful, assumptions, assumptions, normally, standard deviations.
6. Sign test for median, *Calc probability* $= (1/2)^9 = 0.002 < 0.025$. Median significantly different from 1 kg. Agrees with *t*-test conclusion of mean significantly different from 1 kg (mean and median are equal for a normal distribution).
7. The number of cigarettes with nicotine content greater than 0.30 mg.

8. *Calc* $T = 3$, *Tab* $T = 0$. Median of amount of dye washed out for old dye not significantly less than for new dye.
9. (a) Wilcoxon, *Calc* $T = 1.5$, *Tab* $T = 2$. Difference is significant.
   (b) Mann–Whitney U, *Calc* $U = 26.5$, *Tab* $U = 13$. Difference is not significant.
10. Mann–Whitney, *Calc* $U = 46$, *Tab* $U = 23$. Compounds equally effective.
11. Mann–Whitney, *Calc* $U = 9$, *Tab* $U = 37$. Significant difference between brands.

WORKSHEET 12 (Solutions)

1. Numerical.
2. Contingency, individuals, observed.
3. Independent.
4. Expected, E = row total × column total/grand total.
5. 5,

$$\sum \frac{(O - E)^2}{E}$$

except for a 2 × 2 table when we calculate

$$\sum \frac{(|O - E| - \frac{1}{2})^2}{E}$$

rejection.

6. $(r - 1)(c - 1)$, 1.
7. 3.84.
8. *Calc* $\chi^2 = 2.91$, *Tab* $\chi^2 = 3.84$. Proportion of burnt-out tubes for rented sets is not significantly different from proportion for bought sets.
9. *Calc* $\chi^2 = 10.34$, *Tab* $\chi^2 = 7.82$. Proportion of male to female is significantly different for the four garages. Tendency for females to prefer garage B.
10. Combine rows '30–70%' and 'under 30%' to form a 2 × 2 table (low E values). *Calc* $\chi^2 = 1.52$. *Tab* $\chi^2 = 3.84$. The chance of passing is independent of attendance (as defined by $>70\%$ or $\leqslant 70\%$).
11. *Calc* $\chi^2 = 5.48$, *Tab* $\chi^2 = 5.99$. Proportions of A, B, C not significantly different.
12. Fisher exact test, since some E values are below 5 and we have a 2 × 2 table, $a = 4$, $b = 1$, $c = 9$, $d = 6$. Probability = 0.3228, so do not reject the null hypothesis of independence ($0.3228 > 0.05$). (For the really good student, the probabilities of the other 5 tables having the same marginal totals, but with a = 5, 3, 2, 1, 0, respectively, are 0.0830, 0.3874, 0.1761, 0.0293, 0.0014. This

gives a total probability of 0.6126 for a two-sided H_1, or 0.4058 for a one-sided H_1. In both cases H_0 is not rejected.)

13. The standard χ^2 test gives *Calc* $\chi_2^2 = 18.31$, while the trend test gives *Calc* $\chi_1^2 = 14.82$.
    Since $14.82 > 3.84$, there is evidence of a linear trend. Since *Calc* $\chi_3^2 = 18.31 - 14.82 = 3.49$, which is less than 3.84, there is some evidence, but not significant, of some departure from linearity. The proportions across the three categories are 64%, 64% and 76%, respectively (had the middle value been 70% this would have indicated a perfect linear relationship). It shows that, of those getting better, a higher proportion received drug A than was the case with the other two categories of patient.

WORKSHEET 13 (Solutions)

1. Scatter diagram.
2. Correlation coefficient.
3. Correlation coefficient, Pearson, r, ρ.
4. r, arbitrary.
5. -1 to $+1$, negative, zero.
6. ρ, normally distributed, uncorrelated.
7. If x is the percentage increase in unemployment, y the percentage increase in manufacturing output,

   $$\Sigma x = 70, \quad \Sigma x^2 = 1\,136, \quad \Sigma y = -50, \quad \Sigma y^2 = 664, \quad \Sigma xy = -732$$

   $n = 10$, $r = -0.739$, *Calc* $t = -3.10$, *Tab* $t = 1.86$. There is a significant negative correlation, assuming normality of x and y which may be reasonable for these data (see section 7.2). Difficult to draw other conclusions, except that both effects may be due to one or more other factors, such as world slump. In each country there may be individual factors as well, such as lack of investment, government policy, welfare state, attitudes to work and so on.
8. Number of times commercial shown is not normal. Use Spearman's r_s which is 0.736 for these data (using method of Section 13.7, because of ties). *Tab* $r_s = 0.643$, showing significant positive correlation. Increase in number of times commercial shown associated with increase in receipts. Scatter diagram gives the impression that the effect is flattening off after about 30 commercials in the month.
9. No evidence of non-normality or normality. Safer with Spearman's r_s, which is 0.468 (note ties) for these data. *Tab* $r_s = 0.714$. Larger areas are not significantly associated with longer river lengths.
10. No evidence of non-normality or normality. $r_s = -0.964$ (no ties), *Tab* $r_s = 0.714$. Lower death rate is significantly associated with higher percentage using filtered water. Lower death rate could be due to other factors such as

public awareness of the need to boil unfiltered water, and better treatment of typhoid, and so on.

11. Dotplots for income and savings indicate that, while the distribution of income is reasonably normal, the distribution of savings may not be because of positive skewness. Safer to use Spearman's r_s here. $r_s = -0.087$ (since ties), *Tab* $r_s = 0.377$, indicating no significant correlation between income and savings. There are two outliers in the scatter diagram but the effect of leaving them out (which we should not do without a good reason) would probably make no difference to the conclusion above.

12. The variables are 'height' and 'distance' (from home to Oxford). Dotplots show reasonable normality for height but positive skewness for distance. Because of ties use the method of section 13.7 to obtain r_s, which gives $r_s =$ 0.009. Since *Tab* $r_s = 0.648$, this is clearly not significant (as expected).

WORKSHEET 14 (Solutions)

1. Predict, y, x.

2. Scatter, regression, straight line.

3. Did you beat 4000?

4. $\Sigma x = 229$, $\Sigma x^2 = 5333$, $\Sigma y = 3700$, $\Sigma y^2 = 1\,390\,200$, $\Sigma xy = 85\,790$, $n = 10$, $b = 11.92$, $a = 97.0$. Predict (a) 335, (b) 395, (c) 455. Last is least reliable being furthest from the mean value of x, which is 22.9. Also 30 is just outside range of x values in the data.

5. $\Sigma x = 15.05$, $\Sigma x^2 = 55.0377$, $\Sigma y = 15.2$, $\Sigma y^2 = 55.04$, $\Sigma xy = 55.01$, $n = 6$, $b =$ 0.9766, $a = 0.0836$. At zero, predict true depth of 0.08, $s_r = 0.1068$, 95% confidence interval for true depth at zero is -0.13 to 0.30, but negative depths are impossible, so quote 0 to 0.30.

6. $\Sigma x = 289$, $\Sigma x^2 = 12\,629$, $\Sigma y = 73$, $\Sigma y^2 = 839$, $\Sigma xy = 3234$, $n = 7$, $b = 0.316$, $a = -2.60$. At (a) $x = 0$, cannot use equation, (b) $x = 50$, 13.2%, (c) $x =$ 100, cannot use equation. $s_r = 1.28$, *Calc* $t = 6.51$, *Tab* $t = 2.57$, slope is significantly different from zero.

7. $\Sigma x = 965$, $\Sigma x^2 = 78\,975$, $\Sigma y = 370$, $\Sigma y^2 = 11\,598$, $\Sigma xy = 30\,220$, $n = 12$, $b =$ 0.339, $a = 3.55$. (a) at 65, 23.5 to 27.7, (b) at 80, 29.5 to 31.9, (c) at 95, 33.8 to 37.8, using $s_r = 1.78$.

8. $\Sigma x = 55$, $\Sigma x^2 = 385$, $\Sigma y = 740$, $\Sigma y^2 = 57\,150$, $\Sigma xy = 3655$, $n = 10$, $b =$ -5.03, $a = 101.7$, $s_r = 6.15$, *Calc* $t = -7.43$, *Tab* $t = 1.86$, slope is significantly less than zero. 95% confidence interval for β is -6.59 to -3.47.

9. $\Sigma x = 280$, $\Sigma x^2 = 14\,000$, $\Sigma y = 580$, $\Sigma y^2 = 43\,180.44$, $\Sigma xy = 22\,477$, $n = 8$, $b =$ 0.518, $a = 54.4$. (a) 67.3, (b) 82.9, (c) 98.4, but this is extrapolation.

10. Scatter diagram shows no clear pattern, might expect negative correlation. In fact $r = -0.066$, which is not significant (*Calc* $t = -0.19$).

11. Scatter diagram of y against $1/x$ looks linear. Let $z = 1/x$, then (to 2 dps) $\Sigma z = 8.71$, $\Sigma z^2 = 14.53$, $\Sigma y = 44.5$, $\Sigma y^2 = 313.55$, $\Sigma zy = 44.81$, $n = 7$. For $y = a + bz$, $b = -2.86$, $a = 9.92$. For $x = 1$, $z = 1$, predict $y = 7.06$, 95% confidence interval is 6.76 to 7.37 ($s_r = 0.30$). Pearson's r for y and $1/x$ should be larger in magnitude but negative.

WORKSHEET 15 (Solutions)

1. *Calc* $\chi^2 = 0.47$, *Tab* $\chi^2 = 7.82$. Data consistent with theory.
2. (a) *Calc* $\chi^2 = 12.7$, *Tab* $\chi^2 = 11.1$, reject uniform distribution.
   (b) *Calc* $\chi^2 = 60.4$, *Tab* $\chi^2 = 11.1$, reject 2:1:5:4:5:3 distribution. Allowing for volume, significantly more accidents occur during hours of darkness than expected.
3. *Calc* $\chi^2 = 13.7$, *Tab* $\chi^2 = 5.99$. Reject independence, more farms on flat land, fewer on mountainous land than expected.
4. *Calc* $\chi^2 = 5.5$, *Tab* $\chi^2 = 5.99$. Data consistent with 1:2:1 hypothesis.
5. (a) *Calc* $\chi^2 = 9.5$, *Tab* $\chi^2 = 7.82$. Data not consistent with $B(3, 0.5)$ distribution.
   (b) *Calc* $\chi^2 = 3.8$, *Tab* $\chi^2 = 5.99$. Data consistent with $B(3, 0.4633)$ distribution, indicating significantly fewer boys than girls.
6. See section 8.3 for grid method of selecting random points at which to place quadrats. Number the pebbles within a quadrat, and choose ten using random number tables.
   Calc $\chi^2 = 1.9$, *Tab* $\chi^2 = 12.6$. Data consistent with $B(10, 0.4)$ distribution.
7. *Calc* $\chi^2 = 5.5$, *Tab* $\chi^2 = 11.1$. Data consistent with $B(20, 0.1)$ distribution.
8. See section 8.3 for grid method of selecting 80 random points. Estimated m is 1.8, *Calc* $\chi^2 = 16.6$, *Tab* $\chi^2 = 7.82$. Data not consistent with random distribution.
9. *Calc* $\chi^2 = 3.9$, *Tab* $\chi^2 = 9.49$. Data consistent with a random distribution ($m = 2$).
10. (a) $\bar{x} = 2$, $s^2 = 4.13$, so not approximately equal, so not reasonable to assume Poisson.
    (b) *Calc* $\chi^2 = 95.3$, *Tab* $\chi^2 = 7.82$. Data not consistent with random distribution. Many more cars than expected with either no defects or at least 4 defects.
11. (a) $b = 25.11$, *Calc* $W = 0.973$, *Tab* $W = 0.842$, data support normality.
    (b) $b = 18.41$, *Calc* $W = 0.840$, *Tab* $W = 0.829$, data support normality.

MULIPLE CHOICE TEST (Solutions)

1.a	2.a	3.b	4.a	5.b
6.a	7.c	8.c	9.c	10.b

11.b	12.b	13.c	14.a	15.a
16.b	17.c	18.a	19.b	20.a
21.a	22.a	23.a	24.c	25.a
26.a	27.b	28.b	29.b	30.c
31.b	32.c	33.b	34.c	35.b
36.c	37.b	38.a	39.b	40.c
41.c	42.a	43.b	44.c	45.c
46.c	47.a	48.b	49.c	50.b

Appendix D

Statistical tables

ACKNOWLEDGEMENTS

The author would like to thank the following authors and publishers for their kind permission to adapt from the following tables:

Tables D.3, D.5 and D.6 from:

Pearson, E.S. and Hartley, H.O. (1966). *Biometrika Tables for Statisticians*, Vol. 1, 3rd edition, Cambridge University Press, Cambridge.

Table D.7 from:

Runyon, R.P. and Haber, A. (1968). *Fundamentals of Behavioural Statistics*, Addison-Wesley, Reading, Mass.; based on values in Wilcoxon, F., Katti, S. and Wilcox, R.A. (1963), *Critical Values and Probability Levels for the Wilcoxon Rank Sum Test and the Wilcoxon Signed Rank Test*, American Cyanamid Co., New York; and Wilcoxon, F. and Wilcox, R.A. (1964), *Some Rapid Approximate Statistical Procedures*, American Cyanamid Co., New York.

Table D.8 from:

Owen, D.B. (1962), *Handbook of Statistical Tables*, Addison-Wesley, Reading, Mass.; based on values in Auble, D. (1953), Extended tables for the Mann–

Whitney statistic, *Bulletin of the Institute of Educational Research at Indiana University*, 1, 2.

Table D.9 from:

Mead, R. and Curnow, R.N. (1983), *Statistical Methods in Agriculture and Experimental Biology*, Chapman and Hall, London.

Table D.10 from:

Runyon, R.P. and Haber, A. (1968), *Fundamentals of Behavioural Statistics*, Addison-Wesley, Reading, Mass.; based on values in Olds, E.G. (1949), The 5% significance levels of sums of squares of rank differences and a correction, *Annals of Mathematical Statistics*, **20**, 117–18, and Olds, E.G. (1938), Distribution of the sum of squares of rank differences for small numbers of individuals, *Annals of Mathematical Statistics*, **9**, 133–48.

Tables D.11 and 12 from:

Shapiro, S.S. and Wilk, M.B. (1965), An analysis of variance test, for normality and complete samples, *Biometrika*, **52**, 592–611.

TABLE D.1 CUMULATIVE BINOMIAL PROBABILITIES

The table gives the probability of obtaining r or fewer successes in n independent trials, where p = probability of successes in a single trial.

	p =	0.01	0.02	0.03	0.04	0.05	0.06	0.07	0.08	0.09
$n = 2$	$r = 0$	.9801	.9604	.9409	.9216	.9025	.8836	.8649	.8464	.8281
	1	.9999	.9996	.9991	.9984	.9975	.9964	.9951	.9936	.9919
	2	1.0000	1.0000	1.0000	1.0000	1.0000	1.0000	1.0000	1.0000	1.0000
$n = 5$	$r = 0$	.9510	.9039	.8587	.8154	.7738	.7339	.6957	.6591	.6240
	1	.9990	.9962	.9915	.9852	.9774	.9681	.9575	.9456	.9326
	2	1.0000	.9999	.9997	.9994	.9988	.9980	.9969	.9955	.9937
	3		1.0000	1.0000	1.0000	1.0000	.9999	.9999	.9998	.9997
	4						1.0000	1.0000	1.0000	1.0000
$n = 10$	$r = 0$	.9044	.8171	.7374	.6648	.5987	.5386	.4840	.4344	.3894
	1	.9957	.9838	.9655	.9418	.9139	.8824	.8483	.8121	.7746
	2	.9999	.9991	.9972	.9938	.9885	.9812	.9717	.9599	.9460
	3	1.0000	1.0000	.9999	.9996	.9990	.9980	.9964	.9942	.9912
	4			1.0000	1.0000	.9999	.9998	.9997	.9994	.9990
	5					1.0000	1.0000	1.0000	1.0000	.9999
	6									1.0000
$n = 20$	$r = 0$	.8179	.6676	.5438	.4420	.3585	.2901	.2342	.1887	.1516
	1	.9831	.9401	.8802	.8103	.7358	.6605	.5869	.5169	.4516
	2	.9990	.9929	.9790	.9561	.9245	.8850	.8390	.7879	.7334
	3	1.0000	.9994	.9973	.9926	.9841	.9710	.9529	.9294	.9007
	4		1.0000	.9997	.9990	.9974	.9944	.9893	.9817	.9710
	5			1.0000	.9999	.9997	.9991	.9981	.9962	.9932
	6				1.0000	1.0000	.9999	.9997	.9994	.9987
	7						1.0000	1.0000	.9999	.9998
	8								1.0000	1.0000

Continued

$p =$		0.01	0.02	0.03	0.04	0.05	0.06	0.07	0.08	0.09
$n = 50$	$r = 0$	.6050	.3642	.2181	.1299	.0769	.0453	.0266	.0155	.0090
	1	.9106	.7358	.5553	.4005	.2794	.1900	.1265	.0827	.0532
	2	.9862	.9216	.8108	.6767	.5405	.4162	.3108	.2260	.1605
	3	.9984	.9822	.9372	.8609	.7604	.6473	.5327	.4253	.3303
	4	.9999	.9968	.9832	.9510	.8964	.8206	.7290	.6290	.5277
	5	1.0000	.9995	.9963	.9856	.9622	.9224	.8650	.7919	.7072
	6		.9999	.9993	.9964	.9882	.9711	.9417	.8981	.8404
	7		1.0000	.9999	.9992	.9968	.9906	.9780	.9562	.9232
	8			1.0000	.9999	.9992	.9973	.9927	.9833	.9672
	9				1.0000	.9998	.9993	.9978	.9944	.9875
	10					1.0000	.9998	.9994	.9983	.9957
	11						1.0000	.9999	.9995	.9987
	12							1.0000	.9999	.9996
	13								1.0000	.9999
	14									1.0000

Continued

	$p =$	0.10	0.15	0.20	0.25	0.30	0.35	0.40	0.45	0.50
$n = 2$	$r = 0$	.8100	.7225	.6400	.5625	.4900	.4225	.3600	.3025	.2500
	1	.9900	.9775	.9600	.9375	.9100	.8775	.8400	.7975	.7500
	2	1.0000	1.0000	1.0000	1.0000	1.0000	1.0000	1.0000	1.0000	1.0000
$n = 5$	$r = 0$	.5905	.4437	.3277	.2373	.1681	.1160	.0778	.0503	.0313
	1	.9185	.8352	.7373	.6328	.5282	.4284	.3370	.2562	.1875
	2	.9914	.9734	.9421	.8965	.8369	.7648	.6826	.5831	.5000
	3	.9995	.9978	.9933	.9844	.9692	.9460	.9130	.8688	.8125
	4	1.0000	.9999	.9997	.9990	.9976	.9947	.9898	.9815	.9688
	5		1.0000	1.0000	1.0000	1.0000	1.0000	1.0000	1.0000	1.0000
$n = 10$	$r = 0$	.3487	.1969	.1074	.0563	.0282	.0135	.0060	.0025	.0010
	1	.7361	.5443	.3758	.2440	.1493	.0860	.0464	.0233	.0107
	2	.9298	.8202	.6778	.5256	.3828	.2616	.1673	.0996	.0547
	3	.9872	.9500	.8791	.7759	.6496	.5138	.3823	.2660	.1719
	4	.9984	.9901	.9672	.9219	.8497	.7515	.6331	.5044	.3770
	5	.9999	.9986	.9936	.9803	.9527	.9051	.8338	.7384	.6230
	6	1.0000	.9999	.9991	.9965	.9894	.9740	.9452	.8980	.8281
	7		1.0000	.9999	.9996	.9984	.9952	.9877	.9726	.9453
	8			1.0000	1.0000	.9999	.9995	.9983	.9955	.9893
	9					1.0000	1.0000	.9999	.9997	.9990
	10							1.0000	1.0000	1.0000
$n = 20$	$r = 0$	.1216	.0388	.0115	.0032	.0008	.0002			
	1	.3917	.1756	.0692	.0243	.0076	.0021	.0005	.0001	
	2	.6769	.4049	.2061	.0913	.0355	.0121	.0036	.0009	.0002
	3	.8670	.6477	.4114	.2252	.1071	.0444	.0160	.0049	.0013
	4	.9568	.8298	.6296	.4148	.2375	.1182	.0510	.0189	.0059
	5	.9887	.9327	.8042	.6172	.4164	.2454	.1256	.0553	.0207
	6	.9976	.9781	.9133	.7858	.6080	.4166	.2500	.1299	.0577
	7	.9996	.9941	.9679	.8982	.7723	.6010	.4159	.2520	.1316
	8	.9999	.9987	.9900	.9591	.8867	.7624	.5956	.4143	.2517
	9	1.0000	.9998	.9974	.9861	.9520	.8782	.7553	.5914	.4119
	10		1.0000	.9994	.9961	.9829	.9468	.8725	.7507	.5881
	11			.9999	.9991	.9949	.9804	.9435	.8692	.7483
	12			1.0000	.9998	.9987	.9940	.9790	.9420	.8684
	13				1.0000	.9997	.9985	.9935	.9786	.9423
	14					1.0000	.9997	.9984	.9936	.9793
	15						1.0000	.9997	.9985	.9941
	16							1.0000	.9997	.9987
	17								1.0000	.9998
	18									1.0000

Continued

$p =$		0.10	0.15	0.20	0.25	0.30	0.35	0.40	0.45	0.50
$n = 50$	$r = 0$	.0052	.0003							
	1	.0338	.0029	.0002						
	2	.1117	.0142	.0013	.0001					
	3	.2503	.0460	.0057	.0005					
	4	.4312	.1121	.0185	.0021	.0002				
	5	.6161	.2194	.0480	.0070	.0007	.0001			
	6	.7702	.3613	.1034	.0194	.0025	.0002			
	7	.8779	.5188	.1904	.0453	.0073	.0008	.0001		
	8	.9421	.6681	.3073	.0916	.0183	.0025	.0002		
	9	.9755	.7911	.4437	.1637	.0402	.0067	.0008	.0001	
	10	.9906	.8801	.5836	.2622	.0789	.0160	.0022	.0002	
	11	.9968	.9372	.7107	.3816	.1390	.0342	.0057	.0006	
	12	.9990	.9699	.8139	.5110	.2229	.0661	.0133	.0018	.0002
	13	.9997	.9868	.8894	.6370	.3279	.1163	.0280	.0045	.0005
	14	.9999	.9947	.9393	.7481	.4468	.1878	.0540	.0104	.0013
	15	1.0000	.9981	.9692	.8369	.5692	.2801	.0955	.0220	.0033
	16		.9993	.9856	.9017	.6839	.3889	.1561	.0427	.0077
	17		.9998	.9937	.9449	.7822	.5060	.2369	.0765	.0164
	18		.9999	.9975	.9713	.8594	.6216	.3356	.1273	.0325
	19		1.0000	.9991	.9861	.9152	.7264	.4465	.1974	.0595
	20			.9997	.9937	.9522	.8139	.5610	.2862	.1013
	21			.9999	.9974	.9749	.8813	.6701	.3900	.1611
	22			1.0000	.9990	.9877	.9290	.7660	.5019	.2399
	23				.9996	.9944	.9604	.8438	.6134	.3359
	24				.9999	.9976	.9793	.9022	.7160	.4439
	25				1.0000	.9991	.9900	.9427	.8034	.5561
	26					.9997	.9955	.9686	.8721	.6641
	27					.9999	.9981	.9840	.9220	.7601
	28					1.0000	.9993	.9924	.9556	.8389
	29						.9997	.9966	.9765	.8987
	30						.9999	.9986	.9884	.9405
	31						1.0000	.9995	.9947	.9675
	32							.9998	.9978	.9836
	33							.9999	.9991	.9923
	34							1.0000	.9997	.9967
	35								.9999	.9987
	36								1.0000	.9995
	37									.9998
	38									1.0000

Continued

TABLE D.2 CUMULATIVE POISSON PROBABILITIES

The table gives the probability of r or fewer random events per unit time or space, when the average number of such events is m.

m =	0.1	0.2	0.3	0.4	0.5	0.6	0.7	0.8	0.9	1.0
r = 0	.9048	.8187	.7408	.6703	.6065	.5488	.4966	.4493	.4066	.3679
1	.9953	.9825	.9631	.9384	.9098	.8781	.8442	.8088	.7725	.7358
2	.9998	.9989	.9964	.9921	.9856	.9769	.9659	.9526	.9371	.9197
3	1.0000	.9999	.9997	.9992	.9982	.9966	.9942	.9909	.9865	.9810
4		1.0000	1.0000	.9999	.9998	.9996	.9992	.9986	.9977	.9963
5				1.0000	1.0000	1.0000	.9999	.9998	.9997	.9994
6							1.0000	1.0000	1.0000	.9999
7										1.0000

m =	1.1	1.2	1.3	1.4	1.5	1.6	1.7	1.8	1.9	2.0
r = 0	.3329	.3012	.2725	.2466	.2231	.2019	.1827	.1653	.1496	.1353
1	.6990	.6626	.6268	.5918	.5578	.5249	.4932	.4628	.4337	.4060
2	.9004	.8795	.8571	.8335	.8088	.7834	.7572	.7306	.7037	.6767
3	.9743	.9662	.9569	.9463	.9344	.9212	.9068	.8913	.8747	.8571
4	.9946	.9923	.9893	.9857	.9814	.9763	.9704	.9636	.9559	.9473
5	.9990	.9985	.9978	.9968	.9955	.9940	.9920	.9896	.9868	.9834
6	.9999	.9997	.9996	.9994	.9991	.9987	.9981	.9974	.9966	.9955
7	1.0000	1.0000	.9999	.9999	.9998	.9997	.9996	.9994	.9992	.9989
8			1.0000	1.0000	1.0000	1.0000	.9999	.9999	.9998	.9998
9							1.0000	1.0000	1.0000	1.0000

m =	2.1	2.2	2.3	2.4	2.5	2.6	2.7	2.8	2.9	3.0
r = 0	.1225	.1108	.1003	.0907	.0821	.0743	.0672	.0608	.0550	.0498
1	.3796	.3546	.3309	.3084	.2873	.2674	.2487	.2311	.2146	.1991
2	.6496	.6227	.5960	.5697	.5438	.5184	.4936	.4695	.4460	.4232
3	.8386	.8194	.7993	.7787	.7576	.7360	.7141	.6919	.6696	.6472
4	.9379	.9275	.9162	.9041	.8912	.8774	.8629	.8477	.8318	.8153
5	.9796	.9751	.9700	.9643	.9580	.9510	.9433	.9349	.9258	.9161
6	.9941	.9925	.9906	.9884	.9858	.9828	.9794	.9756	.9713	.9665
7	.9985	.9980	.9974	.9967	.9958	.9947	.9934	.9919	.9901	.9881
8	.9997	.9995	.9994	.9991	.9989	.9985	.9981	.9976	.9969	.9962
9	.9999	.9999	.9999	.9998	.9997	.9996	.9995	.9993	.9991	.9989
10	1.0000	1.0000	1.0000	1.0000	.9999	.9999	.9999	.9998	.9998	.9997
11					1.0000	1.0000	1.0000	1.0000	.9999	.9999
12									1.0000	1.0000

Continued

m =	3.1	3.2	3.3	3.4	3.5	3.6	3.7	3.8	3.9	4.0
r = 0	.0450	.0408	.0369	.0334	.0302	.0273	.0247	.0224	.0202	.0183
1	.1847	.1712	.1586	.1468	.1359	.1257	.1162	.1074	.0992	.0916
2	.4012	.3799	.3594	.3397	.3208	.3027	.2854	.2689	.2531	.2381
3	.6248	.6025	.5803	.5584	.5366	.5152	.4942	.4735	.4532	.4335
4	.7982	.7806	.7626	.7442	.7254	.7064	.6872	.6678	.6484	.6288
5	.9057	.8946	.8829	.8705	.8576	.8441	.8301	.8156	.8006	.7851
6	.9612	.9554	.9490	.9421	.9347	.9267	.9182	.9091	.8995	.8893
7	.9858	.9832	.9802	.9769	.9733	.9692	.9648	.9599	.9546	.9489

m =	3.1	3.2	3.3	3.4	3.5	3.6	3.7	3.8	3.9	4.0
8	.9953	.9943	.9931	.9917	.9901	.9883	.9863	.9840	.9815	.9786
9	.9986	.9982	.9978	.9973	.9967	.9960	.9952	.9942	.9931	.9919
10	.9996	.9995	.9994	.9992	.9990	.9987	.9984	.9981	.9977	.9972
11	.9999	.9999	.9998	.9998	.9997	.9996	.9995	.9994	.9993	.9991
12	1.0000	1.0000	1.0000	.9999	.9999	.9999	.9999	.9998	.9998	.9997
13				1.0000	1.0000	1.0000	1.0000	1.0000	.9999	.9999
14									1.0000	1.0000

m =	4.1	4.2	4.3	4.4	4.5	4.6	4.7	4.8	4.9	5.0
r = 0	.0166	.0150	.0136	.0123	.0111	.0101	.0091	.0082	.0074	.0067
1	.0845	.0780	.0719	.0663	.0611	.0563	.0518	.0477	.0439	.0404
2	.2238	.2102	.1974	.1851	.1736	.1626	.1523	.1425	.1333	.1247
3	.4142	.3954	.3772	.3594	.3423	.3257	.3097	.2942	.2793	.2650
4	.6093	.5898	.5704	.5512	.5321	.5132	.4946	.4763	.4582	.4405
5	.7693	.7531	.7367	.7199	.7029	.6858	.6684	.6510	.6335	.6160
6	.8786	.8675	.8558	.8436	.8311	.8180	.8046	.7908	.7767	.7622
7	.9427	.9361	.9290	.9214	.9134	.9049	.8960	.8867	.8769	.8666
8	.9755	.9721	.9683	.9642	.9597	.9549	.9497	.9442	.9382	.9319
9	.9905	.9889	.9871	.9851	.9829	.9805	.9778	.9749	.9717	.9682
10	.9966	.9959	.9952	.9943	.9933	.9922	.9910	.9896	.9880	.9863
11	.9989	.9986	.9983	.9980	.9976	.9971	.9966	.9960	.9953	.9945
12	.9997	.9996	.9995	.9993	.9992	.9990	.9988	.9986	.9983	.9980
13	.9999	.9999	.9998	.9998	.9997	.9997	.9996	.9995	.9994	.9993
14	1.0000	1.0000	1.0000	.9999	.9999	.9999	.9999	.9999	.9998	.9998
15				1.0000	1.0000	1.0000	1.0000	1.0000	.9999	.9999
16									1.0000	1.0000

Continued

m =	5.2	5.4	5.6	5.8	6.0	6.2	6.4	6.6	6.8	7.0
r = 0	.0055	.0045	.0037	.0030	.0025	.0020	.0017	.0014	.0011	.0009
1	.0342	.0289	.0244	.0206	.0174	.0146	.0123	.0103	.0087	.0073
2	.1088	.0948	.0824	.0715	.0620	.0536	.0463	.0400	.0344	.0296
3	.2381	.2133	.1906	.1700	.1512	.1342	.1189	.1052	.0928	.0818
4	.4061	.3733	.3422	.3127	.2851	.2592	.2351	.2127	.1920	.1730
5	.5809	.5461	.5119	.4783	.4457	.4141	.3837	.3547	.3270	.3007
6	.7324	.7017	.6703	.6384	.6063	.5742	.5423	.5108	.4799	.4497
7	.8449	.8217	.7970	.7710	.7440	.7160	.6873	.6581	.6285	.5987
8	.9181	.9027	.8857	.8672	.8472	.8259	.8033	.7796	.7548	.7291
9	.9603	.9512	.9409	.9292	.9161	.9016	.8858	.8686	.8502	.8305
10	.9823	.9775	.9718	.9651	.9574	.9486	.9386	.9274	.9151	.9015
11	.9927	.9904	.9875	.9841	.9799	.9750	.9693	.9627	.9552	.9467
12	.9972	.9962	.9949	.9932	.9912	.9887	.9857	.9821	.9779	.9730
13	.9990	.9986	.9980	.9973	.9964	.9952	.9937	.9920	.9898	.9872
14	.9997	.9995	.9993	.9990	.9986	.9981	.9974	.9966	.9956	.9943
15	.9999	.9998	.9998	.9996	.9995	.9993	.9990	.9986	.9982	.9976
16	1.0000	.9999	.9999	.9999	.9998	.9997	.9996	.9995	.9993	.9990
17		1.0000	1.0000	1.0000	.9999	.9999	.9999	.9998	.9997	.9996

m =	5.2	5.4	5.6	5.8	6.0	6.2	6.4	6.6	6.8	7.0
18					1.0000	1.0000	1.0000	.9999	.9999	.9999
19								1.0000	1.0000	1.0000

m =	7.2	7.4	7.6	7.8	8.0	8.2	8.4	8.6	8.8	9.0
r = 0	.0007	.0006	.0005	.0004	.0003	.0003	.0002	.0002	.0002	.0001
1	.0061	.0051	.0043	.0036	.0030	.0025	.0021	.0018	.0015	.0012
2	.0255	.0219	.0188	.0161	.0138	.0118	.0100	.0086	.0073	.0062
3	.0719	.0632	.0554	.0485	.0424	.0370	.0323	.0281	.0244	.0212
4	.1555	.1395	.1249	.1117	.0996	.0887	.0789	.0701	.0621	.0550
5	.2759	.2526	.2307	.2103	.1912	.1736	.1573	.1422	.1284	.1157
6	.4204	.3920	.3646	.3384	.3134	.2896	.2670	.2457	.2256	.2068
7	.5689	.5393	.5100	.4812	.4530	.4254	.3987	.3728	.3478	.3239
8	.7027	.6757	.6482	.6204	.5925	.5647	.5369	.5094	.4823	.4557
9	.8096	.7877	.7649	.7411	.7166	.6915	.6659	.6400	.6137	.5874
10	.8867	.8707	.8535	.8352	.8159	.7955	.7743	.7522	.7294	.7060
11	.9371	.9265	.9148	.9020	.8881	.8731	.8571	.8400	.8220	.8030
12	.9673	.9609	.9536	.9454	.9362	.9261	.9150	.9029	.8898	.8758
13	.9841	.9805	.9762	.9714	.9658	.9595	.9524	.9445	.9358	.9261
14	.9927	.9908	.9886	.9859	.9827	.9791	.9749	.9701	.9647	.9585
15	.9969	.9959	.9948	.9934	.9918	.9898	.9875	.9848	.9816	.9780
16	.9987	.9983	.9978	.9971	.9963	.9953	.9941	.9926	.9909	.9889
17	.9995	.9993	.9991	.9988	.9984	.9979	.9973	.9966	.9957	.9947
18	.9998	.9997	.9996	.9995	.9993	.9991	.9989	.9985	.9981	.9976
19	.9999	.9999	.9999	.9998	.9997	.9997	.9995	.9994	.9992	.9989
20	1.0000	1.0000	1.0000	.9999	.9999	.9999	.9998	.9998	.9997	.9996
21				1.0000	1.0000	1.0000	.9999	.9999	.9999	.9998
22							1.0000	1.0000	1.0000	.9999
23										1.0000

m =	9.2	9.4	9.6	9.8	10.0	11.0	12.0	13.0	14.0	15.0
$r = 0$	.0001	.0001	.0001	.0001						
1	.0010	.0009	.0007	.0006	.0005	.0002	.0001			
2	.0053	.0045	.0038	.0033	.0028	.0012	.0005	.0002	.0001	
3	.0184	.0160	.0138	.0120	.0103	.0049	.0023	.0011	.0005	.0002
4	.0486	.0429	.0378	.0333	.0293	.0151	.0076	.0037	.0018	.0009
5	.1041	.0935	.0838	.0750	.0671	.0375	.0203	.0107	.0055	.0028
6	.1892	.1727	.1574	.1433	.1301	.0786	.0458	.0259	.0142	.0076
7	.3010	.2792	.2584	.2388	.2202	.1432	.0895	.0540	.0316	.0180
8	.4296	.4042	.3796	.3558	.3328	.2320	.1550	.0998	.0621	.0374
9	.5611	.5349	.5089	.4832	.4579	.3405	.2424	.1658	.1094	.0699
10	.6820	.6576	.6329	.6080	.5830	.4599	.3472	.2517	.1757	.1185
11	.7832	.7626	.7412	.7193	.6968	.5793	.4616	.3532	.2600	.1848
12	.8607	.8448	.8279	.8101	.7916	.6887	.5760	.4631	.3585	.2676
13	.9156	.9042	.8919	.8786	.8645	.7813	.6815	.5730	.4644	.3632
14	.9517	.9441	.9357	.9265	.9165	.8540	.7720	.6751	.5704	.4657
15	.9738	.9691	.9638	.9579	.9513	.9074	.8444	.7636	.6694	.5681

m =	9.2	9.4	9.6	9.8	10.0	11.0	12.0	13.0	14.0	15.0
16	.9865	.9838	.9806	.9770	.9730	.9441	.8987	.8355	.7559	.6641
17	.9934	.9919	.9902	.9881	.9857	.9678	.9370	.8905	.8272	.7489
18	.9969	.9962	.9952	.9941	.9928	.9823	.9626	.9302	.8826	.8195
19	.9986	.9983	.9978	.9972	.9965	.9907	.9787	.9573	.9235	.8752
20	.9994	.9992	.9990	.9987	.9984	.9953	.9884	.9750	.9521	.9170
21	.9998	.9997	.9996	.9995	.9993	.9977	.9939	.9859	.9712	.9469
22	.9999	.9999	.9998	.9998	.9997	.9990	.9970	.9924	.9833	.9673
23	1.0000	1.0000	.9999	.9999	.9999	.9995	.9985	.9960	.9907	.9805
24			1.0000	1.0000	1.0000	.9998	.9993	.9980	.9950	.9888
25						.9999	.9997	.9990	.9974	.9938
26						1.0000	.9999	.9995	.9987	.9967
27							.9999	.9998	.9994	.9983
28							1.0000	.9999	.9997	.9991
29								1.0000	.9999	.9996
30									.9999	.9998
31									1.0000	.9999
32										1.0000

Continued

TABLE D.3(a) NORMAL DISTRIBUTION FUNCTION

For a normal distribution with a mean, μ, and standard deviation, σ, and a particular value of x, calculate $z = (x - \mu)/\sigma$. The table gives the area to the left of x, see Fig. D.1.

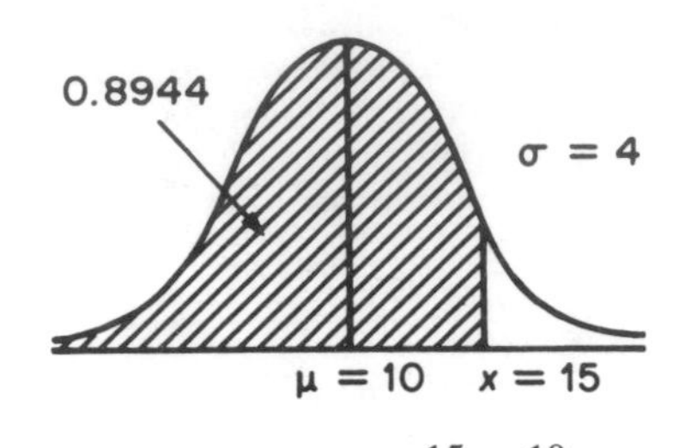

Fig. D.1 $z = \dfrac{x - \mu}{\sigma} = \dfrac{15 - 10}{4} = 1.25.$

$z = \dfrac{(x - \mu)}{\sigma}$	0.00	0.01	0.02	0.03	0.04	0.05	0.06	0.07	0.08	0.09
0.0	.5000	.5040	.5080	.5120	.5160	.5199	.5239	.5279	.5319	.5359
0.1	.5398	.5438	.5478	.5517	.5557	.5596	.5636	.5675	.5714	.5753
0.2	.5793	.5832	.5871	.5910	.5948	.5987	.6026	.6064	.6103	.6141
0.3	.6179	.6217	.6255	.6293	.6331	.6368	.6406	.6443	.6480	.6517
0.4	.6554	.6591	.6628	.6664	.6700	.6736	.6772	.6808	.6844	.6879
0.5	.6915	.6950	.6985	.7019	.7054	.7088	.7123	.7157	.7190	.7224
0.6	.7257	.7291	.7324	.7357	.7389	.7422	.7454	.7486	.7517	.7549
0.7	.7580	.7611	.7642	.7673	.7704	.7734	.7764	.7794	.7823	.7852
0.8	.7881	.7910	.7939	.7967	.7995	.8023	.8051	.8078	.8106	.8133
0.9	.8159	.8186	.8212	.8238	.8264	.8289	.8315	.8340	.8365	.8389
1.0	.8413	.8438	.8461	.8485	.8508	.8531	.8554	.8577	.8599	.8621
1.1	.8643	.8665	.8686	.8708	.8729	.8749	.8770	.8790	.8810	.8830
1.2	.8849	.8869	.8888	.8907	.8925	.8944	.8962	.8980	.8997	.9015
1.3	.9032	.9049	.9066	.9082	.9099	.9115	.9131	.9147	.9162	.9177
1.4	.9192	.9207	.9222	.9236	.9251	.9265	.9279	.9292	.9306	.9319
1.5	.9332	.9345	.9357	.9370	.9382	.9394	.9406	.9418	.9429	.9441
1.6	.9452	.9463	.9474	.9484	.9495	.9505	.9515	.9525	.9535	.9545
1.7	.9554	.9564	.9573	.9582	.9591	.9599	.9608	.9616	.9625	.9633
1.8	.9641	.9649	.9656	.9664	.9671	.9678	.9686	.9693	.9699	.9706
1.9	.9713	.9719	.9726	.9732	.9738	.9744	.9750	.9756	.9761	.9767

Continued

$z = \frac{(x - \mu)}{\sigma}$	0.00	0.01	0.02	0.03	0.04	0.05	0.06	0.07	0.08	0.09
2.0	.9772	.9778	.9783	.9788	.9793	.9798	.9803	.9808	.9812	.9817
2.1	.9821	.9826	.9830	.9834	.9838	.9842	.9846	.9850	.9854	.9857
2.2	.9861	.9864	.9868	.9871	.9875	.9878	.9881	.9884	.9887	.9890
2.3	.9893	.9896	.9898	.9901	.9904	.9906	.9909	.9911	.9913	.9916
2.4	.9918	.9920	.9922	.9925	.9927	.9929	.9931	.9932	.9934	.9936
2.5	.9938	.9940	.9941	.9943	.9945	.9946	.9948	.9949	.9951	.9952
2.6	.9953	.9955	.9956	.9957	.9959	.9960	.9961	.9962	.9963	.9964
2.7	.9965	.9966	.9967	.9968	.9969	.9970	.9971	.9972	.9973	.9974
2.8	.9974	.9975	.9976	.9977	.9977	.9978	.9979	.9979	.9980	.9981
2.9	.9981	.9982	.9982	.9983	.9984	.9984	.9985	.9985	.9986	.9986
3.0	.9987	.9987	.9987	.9988	.9988	.9989	.9989	.9989	.9990	.9990
3.1	.9990	.9991	.9991	.9991	.9992	.9992	.9992	.9992	.9993	.9993
3.2	.9993	.9993	.9994	.9994	.9994	.9994	.9994	.9995	.9995	.9995
3.3	.9995	.9995	.9995	.9996	.9996	.9996	.9996	.9996	.9996	.9997
3.4	.9997	.9997	.9997	.9997	.9997	.9997	.9997	.9997	.9997	.9998

TABLE D.3(b) UPPER PERCENTAGE POINTS FOR THE NORMAL DISTRIBUTION

The table gives the values of z for various right-hand tail areas, α, see Fig. D.2.

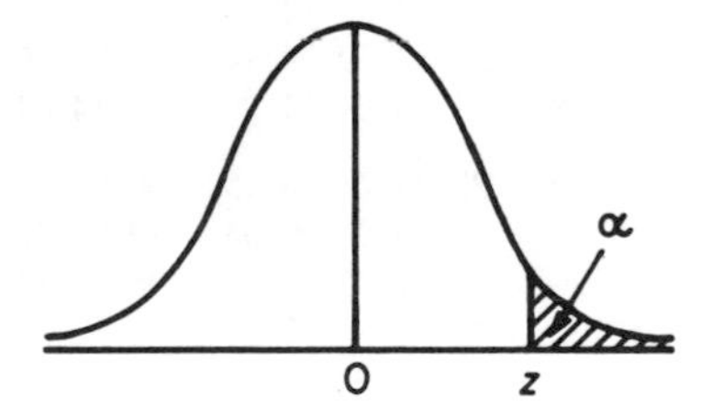

Fig. D.2 Right-hand tail area of a normal distribution.

α	0.05	0.025	0.01	0.005	0.001	0.0005
z	1.645	1.96	2.33	2.58	3.09	3.29

TABLE D.4 RANDOM NUMBERS

30 89 34 43 98	38 51 15 30 26	02 57 93 32 67	19 91 72 23 06	59 24 11 06 50
79 50 49 98 07	05 88 29 05 29	73 15 65 17 92	26 05 21 60 73	55 48 97 54 50
53 64 54 20 36	05 26 90 12 98	73 98 56 47 60	44 54 45 97 21	25 70 96 58 72
87 23 75 21 50	54 47 46 35 72	11 66 30 44 63	69 50 82 74 58	98 25 68 47 79
91 54 58 41 48	70 11 94 79 12	36 63 12 52 72	43 41 11 52 98	91 77 91 85 00
92 41 24 08 42	64 96 82 07 01	40 00 95 09 30	23 40 08 19 78	55 50 92 84 96
65 63 25 34 62	93 01 96 23 23	81 31 94 09 02	75 98 27 85 59	53 09 94 37 37
93 64 13 39 70	98 38 71 77 89	47 98 47 22 09	98 85 91 86 42	30 60 34 07 23
92 44 97 54 10	53 06 50 66 76	13 89 09 41 28	93 04 75 68 09	78 22 82 88 10
69 37 57 14 85	43 72 12 89 80	07 01 17 91 30	17 00 49 53 99	46 51 26 74 28
88 13 45 79 30	32 44 38 84 94	26 65 83 04 43	88 70 99 09 89	31 59 08 29 11
30 86 16 00 13	89 22 16 01 29	98 65 92 13 36	26 88 58 18 89	67 19 71 92 28
19 39 94 95 22	70 99 77 50 29	30 16 69 87 18	48 56 34 92 85	42 54 25 72 84
04 01 90 59 21	33 16 80 53 51	90 02 92 76 72	03 82 77 75 72	33 44 87 58 29
17 45 23 69 94	53 68 59 13 13	68 39 80 62 31	70 44 32 01 47	54 43 70 97 08
13 35 10 58 52	66 73 38 05 80	45 71 76 21 80	10 58 72 17 06	50 72 97 41 48
07 48 12 02 82	51 55 21 61 13	44 27 63 97 04	56 13 88 48 02	34 15 84 30 87
08 16 12 72 05	72 10 63 76 44	92 84 98 81 43	71 66 24 27 16	06 32 39 21 89
51 94 42 32 70	21 82 38 94 46	59 34 75 61 97	72 76 50 50 30	70 27 08 16 72
06 78 72 46 93	36 77 57 19 49	99 18 26 11 63	74 29 96 14 57	76 72 92 86 28
39 14 12 52 96	24 33 70 06 77	56 59 42 11 80	33 05 63 40 14	22 70 62 17 05
71 31 34 36 97	98 57 79 44 68	06 62 74 23 69	77 41 05 17 26	41 68 37 19 53
57 64 15 98 66	13 41 98 06 19	64 53 36 19 16	19 90 71 70 74	04 03 30 05 34
64 26 20 69 40	12 85 65 75 73	92 57 43 97 70	71 28 02 89 91	86 98 64 56 73
91 38 37 54 09	99 35 01 78 03	09 53 57 79 53	50 23 00 90 49	45 28 45 00 94
89 29 45 54 07	22 17 50 32 64	07 30 41 19 36	32 18 08 94 48	20 84 02 47 95
81 31 03 44 27	43 93 91 10 38	72 95 27 58 65	02 23 61 23 17	17 70 26 19 79
05 45 30 21 51	05 14 61 37 61	47 39 50 22 73	28 06 14 72 89	53 64 75 09 70
03 61 43 09 65	35 22 77 22 50	50 37 79 34 14	65 03 56 93 62	34 03 93 18 14
82 75 76 86 14	93 52 73 37 68	83 46 04 11 96	24 14 84 07 19	88 54 05 04 29
62 91 08 18 91	52 65 53 89 39	95 43 21 88 25	36 97 60 89 07	12 03 57 31 39
99 61 53 27 31	18 30 38 21 32	91 03 04 61 53	19 81 45 69 05	35 63 25 00 53
44 29 75 03 84	52 19 73 07 26	92 21 25 48 18	98 14 24 72 12	26 24 89 86 53
51 17 94 61 54	16 39 17 30 32	41 23 37 62 20	51 62 33 79 66	51 95 89 43 55
87 51 27 95 72	31 82 22 31 18	20 31 03 93 60	50 93 18 75 26	62 64 57 46 85
58 12 50 48 30	85 34 65 89 19	63 58 41 42 56	03 67 41 69 48	81 13 44 42 70
78 25 85 91 28	01 85 26 47 58	66 11 84 77 18	30 47 19 42 74	80 13 53 72 66
97 09 87 30 35	04 26 88 10 58	18 44 75 06 52	92 49 73 70 79	49 42 20 09 96
69 08 45 81 37	89 68 51 99 15	33 07 14 39 61	78 05 50 34 14	72 32 78 30 59
82 74 69 78 50	51 47 00 57 40	51 84 26 51 23	14 08 30 96 92	56 71 54 59 96
71 08 26 53 23	43 60 71 41 63	95 26 14 78 09	73 74 63 73 21	06 79 69 81 90
17 60 07 10 21	77 42 60 77 01	20 14 04 09 89	55 79 97 62 57	13 59 38 42 41
90 07 13 82 73	77 37 58 21 35	29 81 98 80 85	51 58 49 82 66	46 94 59 42 25
14 04 16 79 09	72 01 15 51 47	01 12 32 87 84	65 27 89 34 07	40 57 95 06 77
42 44 93 98 30	13 10 61 85 30	46 82 99 79 93	48 62 46 26 71	19 98 34 48 28

TABLE D.5 PERCENTAGE POINTS OF THE *t* DISTRIBUTION

For a t distribution with v degrees of freedom, the table gives the values of t which are exceeded with probability α. Figure D.3 shows a t distribution with $v = 10$ d.f.

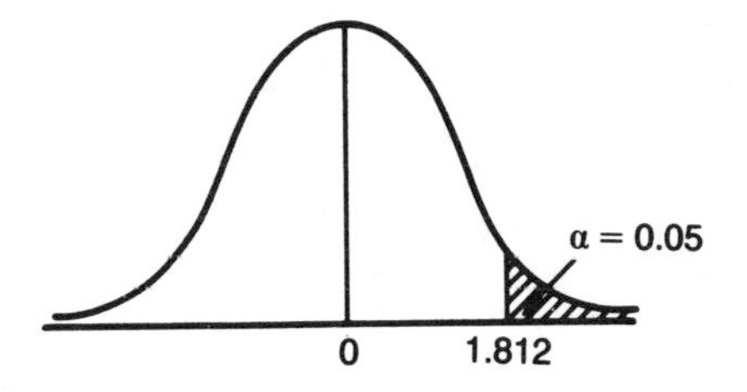

Fig. D.3 t distribution with $v = 10$ d.f.

α =	0.10	0.05	0.025	0.01	0.005	0.001	0.0005
v = 1	3.078	6.314	12.706	31.821	63.657	318.31	636.62
2	1.886	2.920	4.303	6.965	9.925	22.326	31.598
3	1.638	2.353	3.182	4.541	5.841	10.213	12.924
4	1.533	2.132	2.776	3.747	4.604	7.173	8.610
5	1.476	2.015	2.571	3.365	4.032	5.893	6.869
6	1.440	1.943	2.447	3.143	3.707	5.208	5.959
7	1.415	1.895	2.365	2.998	3.499	4.785	5.408
8	1.397	1.860	2.306	2.896	3.355	4.501	5.041
9	1.383	1.833	2.262	2.821	3.250	4.297	4.781
10	1.372	1.812	2.228	2.764	3.169	4.144	4.587
11	1.363	1.796	2.201	2.718	3.106	4.025	4.437
12	1.356	1.782	2.179	2.681	3.055	3.930	4.318
13	1.350	1.771	2.160	2.650	3.012	3.852	4.221
14	1.345	1.761	2.145	2.624	2.977	3.787	4.140
15	1.341	1.753	2.131	2.602	2.947	3.733	4.073
16	1.337	1.746	2.120	2.583	2.921	3.686	4.015
17	1.333	1.740	2.110	2.567	2.898	3.646	3.965
18	1.330	1.734	2.101	2.552	2.878	3.610	3.922
19	1.328	1.729	2.093	2.539	2.861	3.579	3.883
20	1.325	1.725	2.086	2.528	2.845	3.552	3.850
21	1.323	1.721	2.080	2.518	2.831	3.527	3.819
22	1.321	1.717	2.074	2.508	2.819	3.505	3.792
23	1.319	1.714	2.069	2.500	2.807	3.485	3.767
24	1.318	1.711	2.064	2.492	2.797	3.467	3.745
25	1.316	1.708	2.060	2.485	2.787	3.450	3.725
26	1.315	1.706	2.056	2.479	2.779	3.435	3.707

Continued

$\alpha =$	0.10	0.05	0.025	0.01	0.005	0.001	0.0005
27	1.314	1.703	2.052	2.473	2.771	3.421	3.690
28	1.313	1.701	2.048	2.467	2.763	3.408	3.674
29	1.311	1.699	2.045	2.462	2.756	3.396	3.659
30	1.310	1.697	2.042	2.457	2.750	3.385	3.646
40	1.303	1.684	2.021	2.423	2.704	3.307	3.551
60	1.296	1.671	2.000	2.390	2.660	3.232	3.460
120	1.289	1.658	1.980	2.358	2.617	3.160	3.373
∞	1.282	1.645	1.960	2.326	2.576	3.090	3.291

TABLE D.6 5% POINTS OF THE F DISTRIBUTION

The tabulated value is $F_{0.05,\nu_1,\nu_2}$, where $P(X > F_{0.05,\nu_1,\nu_2}) = 0.05$ when X has the F-distribution with ν_1, ν_2 degrees of freedom. The 95% point may be obtained using

$$F_{0.95,\nu_2,\nu_1} = \frac{1}{F_{0.05,\nu_1,\nu_2}}$$

$$\text{e.g.} \quad F_{0.95,12,8} = \frac{11}{F_{0.05,8,12}} = 2.85 = 0.351$$

0.05

0 2.85 $(=F_{0.05,8,12})$

Fig. D.4 F distribution with 8, 12 d.f.

$\nu_1 =$	1	2	3	4	5	6	7	8	10	12	24	∞
$\nu_2 =$ 1	161.4	199.5	215.7	224.6	230.2	234.0	236.8	238.9	241.9	243.9	249.1	254.3
2	18.5	19.0	19.2	19.2	19.3	19.3	19.4	19.4	19.4	19.4	19.5	19.5
3	10.1	9.55	9.28	9.12	9.01	8.94	8.89	8.85	8.79	8.74	8.64	8.53
4	7.71	6.94	6.59	6.39	6.26	6.16	6.09	6.04	5.96	5.91	5.77	5.63
5	6.61	5.79	5.41	5.19	5.05	4.95	4.88	4.82	4.74	4.68	4.53	4.36
6	5.99	5.14	4.76	4.53	4.39	4.28	4.21	4.15	4.06	4.00	3.84	3.67
7	5.59	4.74	4.35	4.12	3.97	3.87	3.79	3.73	3.64	3.57	3.41	3.23
8	5.32	4.46	4.07	3.84	3.69	3.58	3.50	3.44	3.35	3.28	3.12	2.93
9	5.12	4.26	3.86	3.63	3.48	3.37	3.29	3.23	3.14	3.07	2.90	2.71
10	4.96	4.10	3.71	3.48	3.33	3.22	3.14	3.07	2.98	2.91	2.74	2.54
12	4.75	3.89	3.49	3.26	3.11	3.00	2.91	2.85	2.75	2.69	2.51	2.30
15	4.54	3.68	3.29	3.06	2.90	2.79	2.71	2.64	2.54	2.48	2.29	2.07
20	4.35	3.49	3.10	2.87	2.71	2.60	2.51	2.45	2.35	2.28	2.08	1.84
24	4.26	3.40	3.01	2.78	2.62	2.51	2.42	2.36	2.25	2.18	1.98	1.73
30	4.17	3.32	2.92	2.69	2.53	2.42	2.33	2.27	2.16	2.09	1.89	1.62
40	4.08	3.23	2.84	2.61	2.45	2.34	2.25	2.18	2.08	2.00	1.79	1.51
60	4.00	3.15	2.76	2.53	2.37	2.25	2.17	2.10	1.99	1.92	1.70	1.39
∞	3.84	3.00	2.60	2.37	2.21	2.10	2.01	1.94	1.83	1.75	1.52	1.00

TABLE D.7 VALUES OF T FOR THE WILCOXON SIGNED RANK TEST

	Level of significance for one-sided H_1			
	0.05	0.025	0.01	0.005
	Level of significance for two-sided H_1			
n	0.10	0.05	0.02	0.01
5	0	–	–	–
6	2	0	–	–
7	3	2	0	–
8	5	3	1	0
9	8	5	3	1
10	10	8	5	3
11	13	10	7	5
12	17	13	9	7
13	21	17	12	9
14	25	21	15	12
15	30	25	19	15
16	35	29	23	19
17	41	34	27	23
18	47	40	32	27
19	53	46	37	32
20	60	52	43	37
21	67	58	49	42
22	75	65	55	48
23	83	73	62	54
24	91	81	69	61
25	100	89	76	68

TABLE D.8 VALUES OF U FOR THE MANN–WHITNEY U TEST

Critical values of U for the Mann–Whitney test for 0.05 (first value) and 0.01 (second value) significance levels for two-sided H_1, and for 0.025 and 0.005 levels for one-sided H_1.

n_2 \ n_1	1	2	3	4	5	6	7	8	9	10	11	12	13	14	15	16	17	18	19	20
1	–	–	–	–	–	–	–	–	–	–	–	–	–	–	–	–	–	–	–	–
	–	–	–	–	–	–	–	–	–	–	–	–	–	–	–	–	–	–	–	–
2	–	–	–	–	–	–	–	0	0	0	0	1	1	1	1	1	2	2	2	2
	–	–	–	–	–	–	–	–	–	–	–	–	–	–	–	–	–	–	0	0
3	–	–	–	–	0	1	1	2	2	3	3	4	4	5	5	6	6	7	7	8
	–	–	–	–	–	–	–	–	0	0	0	1	1	1	2	2	2	2	3	3
4	–	–	–	0	1	2	3	4	4	5	6	7	8	9	10	11	11	12	13	14
	–	–	–	–	–	0	0	1	1	2	2	3	3	4	5	5	6	6	7	8
5	–	–	0	1	2	3	5	6	7	8	9	11	12	13	14	15	17	18	19	20
	–	–	–	–	0	1	1	2	3	4	5	6	7	7	8	9	10	11	12	13
6	–	–	1	2	3	5	6	8	10	11	13	14	16	17	19	21	22	24	25	27
	–	–	–	0	1	2	3	4	5	6	7	9	10	11	12	13	15	16	17	18
7	–	–	1	3	5	6	8	10	12	14	16	18	20	22	24	26	28	30	32	34
	–	–	–	0	1	3	4	6	7	9	10	12	13	15	16	18	19	21	22	24
8	–	0	2	4	6	8	10	13	15	17	19	22	24	26	29	31	34	36	38	41
	–	–	–	1	2	4	6	7	9	11	13	15	17	18	20	22	24	26	28	30
9	–	0	2	4	7	10	12	15	17	20	23	26	28	31	34	37	39	42	45	48
	–	–	0	1	3	5	7	9	11	13	16	18	20	22	24	27	29	31	33	36
10	–	0	3	5	8	11	14	17	20	23	26	29	33	36	39	42	45	48	52	55
	–	–	0	2	4	6	9	11	13	16	18	21	24	26	29	31	34	37	39	42

Continued

TABLE D.8 *Continued*

n_2 \ n_1	1	2	3	4	5	6	7	8	9	10	11	12	13	14	15	16	17	18	19	20
11	–	0	3	6	9	13	16	19	23	26	30	33	37	40	44	47	51	55	58	62
	–	–	0	2	5	7	10	13	16	18	21	24	27	30	33	36	39	42	45	48
12	–	1	4	7	11	14	18	22	26	29	33	37	41	45	49	53	57	61	65	69
	–	–	1	3	6	9	12	15	18	21	24	27	31	34	37	41	44	47	51	54
13	–	1	4	8	12	16	20	24	28	33	37	41	45	50	54	59	63	67	72	76
	–	–	1	3	7	10	13	17	20	24	27	31	34	38	42	45	49	53	57	60
14	–	1	5	9	13	17	22	26	31	36	40	45	50	55	59	64	69	74	78	83
	–	–	1	4	7	11	15	18	22	26	30	34	38	42	46	50	54	58	63	67
15	–	1	5	10	14	19	24	29	34	39	44	49	54	59	64	70	75	80	85	90
	–	–	2	5	8	12	16	20	24	29	33	37	42	46	51	55	60	64	69	73
16	–	1	6	11	15	21	26	31	37	42	47	53	59	64	70	75	81	86	92	98
	–	–	2	5	9	13	18	22	27	31	36	41	45	50	55	60	65	70	74	79
17	–	2	6	11	17	22	28	34	39	45	51	57	63	69	75	81	87	93	99	105
	–	–	2	6	10	15	19	24	29	34	39	44	49	54	60	65	70	75	81	86
18	–	2	7	12	18	24	30	36	42	48	55	61	67	74	80	86	93	99	106	112
	–	–	2	6	11	16	21	26	31	37	42	47	53	58	64	70	75	81	87	92
19	–	2	7	13	19	25	32	38	45	52	58	65	72	78	85	92	99	106	113	119
	–	0	3	7	12	17	22	28	33	39	45	51	57	63	69	74	81	87	93	99
20	–	2	8	14	20	27	34	41	48	55	62	69	76	83	90	98	105	112	119	127
	–	0	3	8	13	18	24	30	36	42	48	54	60	67	73	79	86	92	99	105

TABLE D.9 PERCENTAGE POINTS OF THE χ^2 DISTRIBUTION

For a χ^2 distribution with ν degrees of freedom, the table gives the values of χ^2 which are exceeded with probability α. See Fig. D.5.

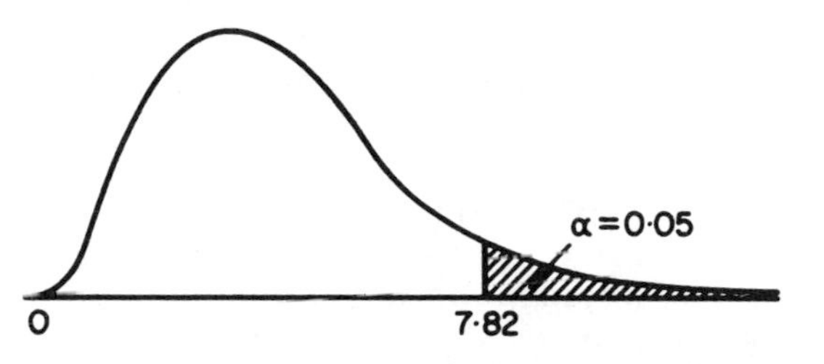

Fig. D.5 χ^2 distribution with $\nu = 3$ d.f.

$\alpha =$	0.50	0.10	0.05	0.025	0.01	0.001
$\nu = 1$	0.45	2.71	3.84	5.02	6.64	10.8
2	1.39	4.61	5.99	7.38	9.21	13.8
3	2.37	6.25	7.82	9.35	11.3	16.3
4	3.36	7.78	9.49	11.1	13.3	18.5
5	4.35	9.24	11.1	12.8	15.1	20.5
6	5.35	10.6	12.6	14.5	16.8	22.5
7	6.35	12.0	14.1	16.0	18.5	24.3
8	7.34	13.4	15.5	17.5	20.1	26.1
9	8.34	14.7	16.9	19.0	21.7	27.9
10	9.34	16.0	18.3	20.5	23.2	29.6
12	11.3	18.5	21.0	23.3	26.2	32.9
15	14.3	22.3	25.0	27.5	30.6	37.7
20	19.3	28.4	31.4	34.2	37.6	45.3
24	23.3	33.2	36.4	39.4	43.0	51.2
30	29.3	40.3	43.8	47.0	50.9	59.7
40	39.3	51.8	55.8	59.3	63.7	73.4
60	59.3	74.4	79.1	83.3	88.4	99.6

TABLE D.10 VALUES OF SPEARMAN'S r_s

	Level of significance for one-sided H_1			
	0.05	0.025	0.01	0.005
	Level of significance for two-sided H_1			
n	0.1	0.05	0.02	0.01
5	0.900	1.000	1.000	–
6	0.829	0.886	0.943	1.000
7	0.714	0.786	0.893	0.929
8	0.643	0.738	0.833	0.881
9	0.600	0.683	0.783	0.833
10	0.564	0.648	0.746	0.794
12	0.506	0.591	0.712	0.777
14	0.456	0.544	0.645	0.715
16	0.425	0.506	0.601	0.665
18	0.399	0.475	0.564	0.625
20	0.377	0.450	0.534	0.591
22	0.359	0.428	0.508	0.562
24	0.343	0.409	0.485	0.537
26	0.329	0.392	0.465	0.515
28	0.317	0.377	0.448	0.496
30	0.306	0.364	0.432	0.478

TABLE D.11 COEFFICIENTS FOR THE SHAPIRO–WILK TEST FOR NORMALITY

$n =$	2	3	4	5	6	7	8	9	10
a_1	0.7071	0.7071	0.6872	0.6646	0.6431	0.6233	0.6052	0.5888	0.5739
a_2	–	.0000	.1677	.2413	.2806	.3031	.3164	.3244	.3291
a_3	–	–	–	.0000	.0875	.1401	.1743	.1976	.2141
a_4	–	–	–	–	–	.0000	.0561	.0947	.1224
a_5	–	–	–	–	–	–	–	.0000	.0399

$n =$	11	12	13	14	15	16	17	18	19	20
a_1	0.5601	0.5475	0.5359	0.5251	0.5150	0.5056	0.4968	0.4886	0.4808	0.4734
a_2	.3315	.3325	.3325	.3318	.3306	.3290	.3273	.3253	.3232	.3211
a_3	.2260	.2347	.2412	.2460	.2495	.2521	.2540	.2553	.2561	.2565
a_4	.1429	.1586	.1707	.1802	.1878	.1939	.1988	.2027	.2059	.2085
a_5	.0695	.0922	.1099	.1240	.1353	.1447	.1524	.1587	.1641	.1686
a_6	0.0000	0.0303	0.0539	0.0727	0.0880	0.1005	0.1109	0.1197	0.1271	0.1334
a_7	–	–	.0000	.0240	.0433	.0593	.0725	.0837	.0932	.1013
a_8	–	–	–	–	.0000	.0196	.0359	.0496	.0612	.0711
a_9	–	–	–	–	–	–	.0000	.0163	.0303	.0422
a_{10}	–	–	–	–	–	–	–	–	.0000	.0140

TABLE D.11 *Continued*

$n =$	21	22	23	24	25	26	27	28	29	30
a_1	0.4643	0.4590	0.4542	0.4493	0.4450	0.4407	0.4366	0.4328	0.4291	0.4254
a_2	.3185	.3156	.3126	.3098	.3069	.3043	.3018	.2992	.2968	.2944
a_3	.2578	.2571	.2563	.2554	.2543	.2533	.2522	.2510	.2499	.2487
a_4	.2119	.2131	.2139	.2145	.2148	.2151	.2152	.2151	.2150	.2148
a_5	.1736	.1764	.1787	.1807	.1822	.1836	.1848	.1857	.1864	.1870
a_6	0.1399	0.1443	0.1480	0.1512	0.1539	0.1563	0.1584	0.1601	0.1616	0.1630
a_7	.1092	.1150	.1201	.1245	.1283	.1316	.1346	.1372	.1395	.1415
a_8	.0804	.0878	.0941	.0997	.1046	.1089	.1128	.1162	.1192	.1219
a_9	.0530	.0618	.0696	.0764	.0823	.0876	.0923	.0965	.1002	.1036
a_{10}	.0263	.0368	.0459	.0539	.0610	.0672	.0728	.0778	.0822	.0862
a_{11}	0.0000	0.0122	0.0228	0.0321	0.0403	0.0476	0.0540	0.0598	0.0650	0.0697
a_{12}	–	–	.0000	.0107	.0200	.0284	.0358	.0424	.0483	.0537
a_{13}	–	–	–	–	.0000	.0094	.0178	.0253	.0320	.0381
a_{14}	–	–	–	–	–	–	.0000	.0084	.0159	.0227
a_{15}	–	–	–	–	–	–	–	–	.0000	.0076

TABLE D.12 PERCENTAGE POINTS OF W FOR THE SHAPIRO–WILK TEST FOR NORMALITY

Level of significance

n	0.01	0.02	0.05	0.10	0.50	0.90	0.95	0.98	0.99
3	0.753	0.756	0.767	0.789	0.959	0.998	0.999	1.000	1.000
4	.687	.707	.748	.792	.935	.987	.992	.996	.997
5	.686	.715	.762	.806	.927	.979	.986	.991	.993
6	0.713	0.743	0.788	0.826	0.927	0.974	0.981	0.986	0.989
7	.730	.760	.803	.838	.928	.972	.979	.985	.988
8	.749	.778	.818	.851	.932	.972	.978	.984	.987
9	.764	.791	.829	.859	.935	.972	.978	.984	.986
10	.781	.806	.842	.869	.938	.972	.978	.983	.986
11	0.792	0.817	0.850	0.876	0.940	0.973	0.979	0.984	0.986
12	.805	.828	.859	.883	.943	.973	.979	.984	.986
13	.814	.837	.866	.889	.945	.974	.979	.984	.986
14	.825	.846	.874	.895	.947	.975	.980	.984	.986
15	.835	.855	.881	.901	.950	.975	.980	.984	.987
16	0.844	0.863	0.887	0.906	0.952	0.976	0.981	0.985	0.987
17	.851	.869	.892	.910	.954	.977	.981	.985	.987
18	.858	.874	.897	.914	.956	.978	.982	.986	.988
19	.863	.879	.901	.917	.957	.978	.982	.986	.988
20	.868	.884	.905	.920	.959	.979	.983	.986	.988
21	0.873	0.888	0.908	0.923	0.960	0.980	0.983	0.987	0.989
22	.878	.892	.911	.926	.961	.980	.984	.987	.989
23	.881	.895	.914	.928	.962	.981	.984	.987	.989
24	.884	.898	.916	.930	.963	.981	.984	.987	.989
25	.888	.901	.918	.931	.964	.981	.985	.988	.989
26	0.891	0.904	0.920	0.933	0.965	0.982	0.985	0.988	0.989
27	.894	.906	.923	.935	.965	.982	.985	.988	.990
28	.896	.908	.924	.936	.966	.982	.985	.988	.990
29	.898	.910	.926	.937	.966	.982	.985	.988	.990
30	.900	.912	.927	.939	.967	.983	.985	.988	.990

Appendix E

Glossary of symbols

The references in brackets refer to the main chapter and section in which the symbol is introduced, defined or used.

ROMAN SYMBOLS

Symbol	Meaning
a	Intercept of sample regression line (14.1)
b	Slope of sample regression line (14.1)
$B(n, p)$	General binomial distribution (6.3)
c	Number of columns in contingency table (12.2)
d	Difference between pairs of values (9.10)
d	Difference between pairs of ranks (13.5)
$\bar{d}$	Sample mean difference (9.10, 10.12)
e	2.718 (2.4, 6.9)
E	Event (5.3)
E	Expected frequency (12.3, 15.2)
E'	Not E, complement of event E (5.12)
H_0	Null hypothesis (10.1)
H_1	Alternative hypothesis (10.1)
m	Mean number of random events per unit time or space (6.9)
n	Number of values in sample (= sample size) (2.1, 9.1, 9.6, 9.9)
n	Total number of equally likely outcomes (5.3)
n	(Large) number of trials (5.3)
n	Number of trials in a binomial experiment (6.3)
n	Number of pairs (9.10, 10.12)
n	Number of individuals in correlation and regression (13.2, 14.2)
$n!$	Factorial n (2.2)

$\binom{n}{x}$	$\dfrac{n!}{x!(n-x)!}$ (6.3)
$N(\mu, \sigma^2)$	General normal distribution (7.4)
O	Observed frequency (12.3, 15.2)
p	Probability of success in a single trial of a binomial experiment (6.3)
$\mathrm{P}(E)$	Probability that event E will occur (5.3)
$\mathrm{P}(E_1$ and $E_2)$	Probability that both events E_1 and E_2 will occur (5.8)
$\mathrm{P}(E_1$ or E_2 or both)	Probability that either or both of events E_1 and E_2 will occur (5.8)
$\mathrm{P}(E_2 \mid E_1)$	Probability that event E_2 will occur, given E_1 has occurred (5.9)
$\mathrm{P}(x)$	Probability of x successes in binomial experiment (6.3)
$\mathrm{P}(x)$	Probability of x random events per unit time or space (6.9)
r	Number of trials resulting in event E (5.3)
r	Number of equally likely outcomes resulting in event E(5.3)
r	Number of rows in contingency table (12.2)
r	Sample value of Pearson's correlation coefficient (13.2)
r_s	Sample value of Spearman's correlation coefficient (13.5)
R_1, R_2	Sum of ranks of values in samples of sizes n_1, n_2 in Mann–Whitney U test (11.9)
s	Sample standard deviation (4.7)
s^2	Sample variance (4.10)
s^2	Pooled estimate of variance (9.11)
s_d	Sample standard deviation of differences (9.10, 10.12)
s_r	Residual standard deviation (14.5)
t	Student's statistic (9.4, 9.10, 10.9)
T	Wilcoxon signed rank statistic (11.5)
T_+, T_-	Sum of ranks of positive and negative differences in Wilcoxon signed rank test (11.6)
U	Mann–Whitney statistic (11.9)
U_1, U_2	Values calculated in Mann–Whitney U test (11.9)
x	Any variable (2.1, 6.3, 6.9, 13.2)
x	Variable used to predict y variable in regression (14.1)
$x_{(1)}$, $x_{(2)}$, $x_{(n)}$, $x_{(i)}$	The first, second, nth, ith, values of x in order of magnitude (15.5)
x_0	A particular value of x in regression (14.5)
$\bar{x}$	Sample mean of x (2.1, 4.2)
y	Variable used with variable x in correlation (13.2)
y	Variable to be predicted from x variable in regression (14.1)
$\bar{y}$	Sample mean of y (14.1)
z	Value used in Table D.3 (areas of normal distribution) (7.3)

GREEK SYMBOLS

α (alpha)	Probability used in t tables (9.4)

α	Probability used in χ^2 tables (12.3)
α	Intercept of population regression line (14.6)
β (beta)	Slope of population regression line (14.6)
μ (mu)	Mean of a normal distribution (7.2)
μ	Population mean (8.5, 9.2, 10.3)
μ_d	Mean of a population of differences (9.10, 10.12)
$\mu_{\bar{x}}$	Mean of distribution of $\bar{x}$ (8.5)
ν (nu)	Degrees of freedom (9.4, 9.7, 10.9, 10.13, 12.3, 13.3)
ρ (rho)	Population value of Pearson's correlation coefficient (13.2)
σ (sigma)	Standard deviation of a normal distribution (7.2)
σ	Population standard deviation (8.5)
$\sigma_{\bar{x}}$	Standard deviation of distribution of $\bar{x}$ (8.5)
Σ (sigma)	Operation of summing (2.1)
χ^2 (chi-squared)	Chi-squared statistic (12.1, 15.1)

Appendix F

Introduction to Minitab data entry and list of Minitab commands

Minitab is an interactive statistical computer package which is widely employed in colleges as an aid to teaching statistics. It has been used in nearly all the chapters of this book, where it was assumed that the reader was familiar with data entry, data editing and some simple commands. The purpose of this appendix is to provide a brief introduction to these aspects of Minitab, using the package in command mode rather than Windows mode.

No attempt has been made to give a detailed account of the package. For such a treatment readers should consult B.F. Ryan, B.L. Joiner and T.A. Ryan, *Minitab Handbook* (2nd edn), PWS-Kent, Boston (1985) or a Minitab reference manual (Minitab Inc., 3081 Enterprise Drive, State College, PA 16801, USA).

Naturally, the package must be loaded into your computer and you must know how to access it, for example by highlighting it on a menu of packages and pressing the RETURN (or ENTER) key, or by typing Minitab followed by RETURN or ENTER. The screen response should be:

MTB>

This implies that you have correctly accessed Minitab and you can now type in commands and data.

When you use Minitab to analyse statistical data, you first have to enter the data into a 'worksheet' consisting of rows and columns. Usually the data for a particular variable are held in a particular column. Columns are referred to as C1, C2, ... If the command SET C1 is typed in and entered, the screen response is

DATA>

and you should now type in the data (for the variable you want to hold in C1) in rows, say 10 numbers to a row, each number being separated from the next by a space. Each row of data should be entered into the computer by pressing the RETURN (or ENTER) key. When you have finished entering the data, you type END after the response DATA>, and once again press the RETURN key.

Example

Enter the numbers 183, 163, 152, 157, 157, 165, 173, 180, 164, 160, 166, and 157 into column C1, name C1 Height, and print out the contents of C1 on the screen. The screen should show the following:

```
MTB> SET C1
DATA> 183 163 152 157 157 165 173 180 164 160
DATA> 166 157
DATA> END
MTB> NAME  C1  'HEIGHT'
MTB> PRINT C1
12 ROWS READ
HEIGHT
183 163 152 157 157 165 173 180 164 160 166 157
```

Note: The two single quotes round the word 'HEIGHT' are entered using the key to the right of the key containing a colon : and a semi-colon ;.

When data for two or more variables are required to be input at the same time, the READ command is useful (as an alternative to the SET command). So we could have used the following to input data for 'SALES' and 'SCORE' in Table 14.2:

```
MTB> READ C1 C2
DATA> 105 45
DATA> 120 75
DATA> 160 85
DATA> 155 65
DATA> 70 50
DATA> 150 70
DATA> 185 80
DATA> 130 55
DATA> END
```

If you make a mistake entering data, you can use the back-arrow (←) key if you notice the mistake before you have entered the current line into the computer. If, however, the line containing the mistake has already been entered, it is best to carry on until you have completed all the data entry. Then edits can be made using one of the following commands: LET, INSERT, DELETE or ERASE. For the 'SALES' and 'SCORE' example above:

```
MTB> LET C2(3) = 90
```

would change the third value in C2 from 85 to 90.

```
MTB> INSERT  6  7  C1
DATA> 160
DATA> END
```

would put 160 in C1 between the 6th value (i.e. 150) and the 7th value (i.e. 185). Column C1 would then have 9 rows instead of 8.

MTB> DELETE 4 C1

Table F.1 Minitab commands

Command	*Use*	*Section reference*
BOXPLOT	Box and whisker plots	3.2, 4.10
CDF	Probabilities for distributions	7.4, 6.7, 6.15
CHISQUARE	χ^2 test	12.5
CORRELATION	Pearson's correlation coefficient	13.8
DELETE	Deleting a row of data	App. F
DESCRIBE	Summary statistics	App. F
DOTPLOT	Horizontal plot for one variable	3.2, 3.11
END	End of a data list	3.2, App. F
ERASE	Erasing a column of data	App. F
HISTOGRAM	Histogram for one variable	3.2
INSERT	Insert a row of data	App. F
INVCDF	Inverse of CDF, for normal distribution	7.4
LET	Editing data, creating new variables	App. F, 9.10
MANNWHITNEY	Mann–Whitney U test	11.11
NAME	Giving a name to a column variable	App. F, 3.2
NOOUTFILE	Ending an output file (for hard copy)	3.2
OUTFILE	Creating an output file (for hard copy)	3.2
PDF	Binomial and Poisson probabilities	6.7, 6.15
PLOT	Scatter diagram	3.4
PRINT	Printing data on screen	App. F, 3.2
RANDOM	Generating random numbers	7.5, 8.6, 6.8
RANK	Putting data in order of magnitude	13.8
READ	Inputting data in columns, to be stored in same columns	App. F, 3.4
REGRESSION	Regression analysis	14.8
RMEAN	Calculating the mean of a row of data	8.6
SET	Inputting data in rows, to be stored in a column	App. F, 3.2
STEM-AND-LEAF	Stem and leaf plot	3.2
STEST	Sign test	11.5
STOP	To 'leave' Minitab, i.e. to end a Minitab session	3.2
TINTERVAL	Confidence interval for μ, μ_d	9.4, 9.10
TTEST	Hypothesis test for μ, μ_d	10.10, 10.12
TWOSAMPLE-T	Confidence interval and hypothesis test for $(\mu_1 - \mu_2)$	9.11, 10.13
WTEST	Wilcoxon signed rank test	11.8

Table F.2 Minitab subcommands

Sub-command	*Used with command*
ALTERNATIVE	TTEST, STEST, WTEST
BINOMIAL	CDF, PDF, RANDOM
BY	DOTPLOT, BOXPLOT, DESCRIBE
INCREMENT	HISTOGRAM
NORMAL	CDF, INVCDF, RANDOM
POISSON	CDF, PDF, RANDOM
POOLED	TWOSAMPLE-T
PREDICT	REGRESSION
RESIDUALS	REGRESSION
START	HISTOGRAM

would delete the 4th value (i.e. 155) in C1, reducing the number of rows in C1 by 1.

MTB> ERASE C2

would erase all the data in C2.

When you have completed all the edits it is a good idea to type in a print command, for example PRINT C1 C2, to check on the screen that all your data is now correct.

Many Minitab commands have been used in various sections of this book, and have been explained as they arose. One which is generally useful, and is worth discussing here in more detail, is the DESCRIBE command. It was mentioned in sections 4.2, 4.3, 4.7 and 4.8. The command DESCRIBE C1 in fact gives ten statistics about the data held in C1. These statistics are:

N, the number of data items, i.e. the number of rows.
MEAN, the mean of the data (section 4.2).
MEDIAN, the median of the data (section 4.4).
TRMEAN, the trimmed mean, i.e. the mean excluding the smallest 5% and the largest 5% of values (this command is seldom if ever useful).
STDEV, the standard deviation of the data (section 7.4).
SEMEAN, the 'standard error of the mean', i.e. STDEV/$\sqrt{N}$ (not used much in this book).
MIN, MAX, the smallest and largest data values, respectively.
Q_1, Q_3, the lower and upper quartiles, respectively (section 4.8).

Example

Apply the DESCRIBE command to the 12 observed values of 'HEIGHT' given earlier in this appendix:

MTB> DESCRIBE 'HEIGHT'

	N	MEAN	MEDIAN	TRMEAN	STDEV	SEMEAN
HEIGHT	12	164.75	163.5	164.2	9.55	2.76
	MIN	MAX	Q_1	Q_3		
	152	183	157	171.25		

Finally, Tables F.1 and F.2 list the Minitab commands and sub-commands used in this book, together with a section reference for each.

Index